Environmental Applications of Geochemical Modeling

Geochemical modeling is a powerful tool for characterizing environmental site contaminations and predicting environmental impacts. This book discusses the application of geochemical models to environmental practice and studies, through the use of numerous case studies of real-world environmental problems. It will thus help students to relate class-room learning to real-life issues, and help practitioners to see more clearly the benefits of geochemical modeling for their own work.

Environmental Applications of Geochemical Modeling opens with background chapters on thermodynamics, kinetics, and geochemical modeling in general. Case studies are then presented which illustrate the application of various types of geochemical models to such diverse problems as acid mine drainage, nuclear waste disposal, bioavailability and risk assessment, mine tailings and mining waste, pit lake chemistry, deep well injection, chemical weathering, artificial recharge, landfill leachate, and microbial respiration rates. In each example the authors clearly define the environmental threat in question; explain how geochemical modeling may help solve the problem posed; and advise the reader how to prepare input files for geochemical codes and interpret the results in terms of meeting regulatory requirements. Models covered include speciation–solubility, surface complexation, mass transfer/reaction path, inverse mass balance, and coupled reactive transport models. The examples also demonstrate the use of many popular geochemical modeling codes such as PHREEQC, MINTEQA2, EQ3/6, and The Geochemist's Workbench™. Support material for the book, including program codes, input files and exercise problems, is available on the internet.

This book will serve as an advanced textbook for courses in environmental geochemistry, and as an indispensable reference for professional hydrogeologists, geochemists, engineers, and regulators, working in the environmental consulting industry.

CHEN ZHU received a Master's degree from the University of Toronto and a Ph.D. degree from Johns Hopkins University (1992), before being awarded a post-doctoral fellowship at Woods Hole Oceanographic Institution. He subsequently worked for five years in the environmental industry, where he encountered many environmental problems first-hand and routinely used geochemical modeling as a tool in his work. Dr Zhu is now an assistant professor at the University of Pittsburgh where he teaches courses in geochemical modeling.

GREG ANDERSON has been Professor of Geochemistry at the University of Toronto for 35 years and is the author of two textbooks on thermodynamics for Earth scientists: *Thermodynamics in Geochemistry* (1993) and *Thermodynamics of Natural Systems* (1995). In 2000 he was awarded the Past President's medal by The Mineralogical Association of Canada for contributions to geochemistry.

ENVIRONMENTAL APPLICATIONS OF GEOCHEMICAL MODELING

CHEN ZHU
University of Pittsburgh

GREG ANDERSON
University of Toronto

CAMBRIDGE
UNIVERSITY PRESS

PUBLISHED BY THE PRESS SYNDICATE OF THE UNIVERSITY OF CAMBRIDGE
The Pitt Building, Trumpington Street, Cambridge, United Kingdom

CAMBRIDGE UNIVERSITY PRESS
The Edinburgh Building, Cambridge CB2 2RU, UK
40 West 20th Street, New York, NY 10011-4211, USA
477 Williamstown Road, Port Melbourne, VIC 3207, Australia
Ruiz de Alarcón 13, 28014 Madrid, Spain
Dock House, The Waterfront, Cape Town 8001, South Africa

First published 2002

Printed in the United Kingdom at the University Press, Cambridge

Typeface MathTime from Y&Y Inc.　　*System* Y&Y TEX System

A catalogue record for this book is available from the British Library

Library of Congress Cataloguing in Publication data

Zhu, Chen, 1962–
　　Environmental applications of geochemical modeling / Chen Zhu and G.M. Anderson.
　　　　p.　cm.
　　Includes bibliographical references and index.
　　ISBN 0 521 80907 X – ISNB 0 521 00577 9 (pb.)
　　1. Environmental geochemistry – Mathematical models. 2. Geochemical modeling.
I. Anderson, G. M. (Gregor Munro), 1932– II. Title.

　　QE516.4 .Z48 2002
　　551.9–dc212001043002

ISBN 0 521 80907 X　hardback
ISBN 0 521 00577 9　paperback

To Leilei, Agnes, and Gregory

Contents

Preface

This book concerns applications of geochemical modeling to "real-world" environmental problems and issues. What makes the environmental problems and issues real is that federal and state environmental regulations mandate certain actions being taken to prevent, reduce, or remediate environmental pollution. The fact that environmental work is performed within a regulatory framework is essential to the understanding of the role of geochemical modeling and geochemistry in general. It must be understood that environmental work has been, is being, and will be carried out to comply with various regulations every day, regardless of how imperfect our models are or how limited our knowledge might be. Society has a need: human health is at risk, and the environment is deteriorating. Neither situation can afford to wait for scientists to provide all the answers or to wait for agreement before any actions are taken. We think the only course of action is to put the best science available in the hands of environmental specialists.

It is, therefore, an objective of this book to bridge the two worlds, academic research and environmental practice, that seem ever apart. We want to provide a guide on geochemical modeling for practicing geochemists, hydrogeologists, engineers, and regulators. Most practitioners in the environmental field lack formal training in geochemical modeling but have to work under stringent deadline, budget, and regulatory constraints. We hope that this book – with its balanced focus on geochemical background, computer programs, and real-world case studies – will serve as a desktop reference for them.

We also want this book to help students and academics to connect geochemistry with society's environmental problems. A wide variety of environmental issues are covered in this book, and most examples are drawn from real-world cases; some from first-hand experience. We hope this book will fill some gaps in the education of geology students.

We believe that research in environmental geochemistry has a bright future and we hope that this book will convey a sense of this promise. The geochemistry of the subsurface is a very complex subject, no less scientifically challenging than, say, the studies of ore deposits. It will be evident from the case studies in this book that our ability to model geochemical reactions in the near Earth surface environment is limited by a lack of progress on fundamental issues in geochemistry. There is no lack of challenges!

The current "state-of-the-practice" of environmental applications of geochemical modeling is at an interesting conjuncture. On one hand, the role of geochemistry is becoming increasingly recognized lately. The National Research Council has attributed the failure of pump-and-treat remediation at many sites partly to the lack of understanding of geochemistry. On the other hand, geochemical modeling is nowhere near the usefulness or the status of groundwater flow and contaminant transport modeling.

It is a widely held belief, albeit among non-geochemists, that geochemical modeling does not produce practically useful results. We hope that the case studies demonstrate the utility of geochemical modeling and that this book will bolster the applications of geochemical modeling in environmental practice.

Finally, we believe the modeling concepts, techniques, and case studies introduced in this book could be useful for studies of traditional geological topics and processes, such as ore deposits, sediment diagenesis, and metamorphism. In fact, geologists have much to gain by closely examining environmental applications of geochemical modeling because, in shallow active geological systems where contamination occurs, the boundary conditions are better constrained than in fossil geological systems, and samples for both solid and water are obtainable, even years after model predictions. This makes testing geochemical models possible.

We are indebted to a number of individuals whom we want to acknowledge here. Thanks are due to the reviewers: Martin Appold, John Apps, Gordon Bennett, David Dzombak, and David Parkhurst. Their time and interests have improved the quality of this book. This book took us five years, and it could not have been completed without the constant encouragement, interests, and moral support from many friends and colleagues: Neil Coleman, David Dzombak, John Ferry, Bill Ford, English Pearcy, Jeff Raffensperger, Frank Schwartz, Kurt Swingle, David and Linda Veblen. Various assistance from Adam Nagle, Erica Love, Dana McClish, Scott Mest, and Amy Scerba are appreciated. We particularly want to thank David S. Burden of US EPA. David was going to write this book with us, but his increasing responsibilities at the Agency have taken away more and more of his "free time" and prevented him from doing so. However, we thank him for his enthusiasm and interest, and sorrily miss his contribution. CZ also wants to acknowledge Rick Waddell of GeoTrans, Inc. Rick has introduced him to many environmental problems. Funding from the National Science Foundation (EAR0003816) and the US Environmental Protection Agency is appreciated. Lastly, the publication of this book would not be possible without the patience and skills of our editors, Drs Matthew Lloyd and Susan Francis. We cannot thank them enough.

Support material for the book, including program codes, input files and exercise problems, is available on the internet: www.pitt.edu/~czhu/book.html.

1

Introduction

1.1 Environmental Problems and the Need for Geochemical Modeling

In this chapter, we introduce some of the major environmental problems that our societies face, and point out the need for using geochemical modeling.

What kinds of problems are addressed or can be addressed by geochemical modeling? Some important problems are as follows.

1.1.1 High-Level Radioactive Waste Disposal

Nuclear power generation and weapons production have resulted in tons of spent fuel and high-level nuclear wastes. These wastes are currently stored under temporary and deteriorating conditions and are awaiting disposal in a permanent repository. In the USA, the Department of Energy is directed by the Congress to investigate the suitability of Yucca Mountain, southern Nevada, to host a permanent geological repository for high-level nuclear waste (Figure 1.1). It is proposed that the thick unsaturated volcanic tuff in the Nevada desert will be a suitable location for a permanent geologic repository because: (1) a deep groundwater table is 800 to 1 000 feet (ca. 250 to 300 m) below the proposed repository; (2) the site is far from large populations; and (3) a desert climate that has rainfall less than 6 inches (ca. 15 cm) per year limits downward percolation of water and vapor (Figure 1.2).

Before Yucca Mountain can become the nuclear waste repository for the nation, a license must be issued by the US Nuclear Regulatory Commission. Evaluation of the site for safe storage of high-level nuclear wastes for at least 10 000 years requires a broad spectrum of scientific disciplines. Mathematical models are developed to calculate the amount and type of radioactive materials that could be released into the environment due to different processes and events.

Two features of the project make geochemical modeling an indispensable and valuable tool that it has never been before. First, it concerns future events: what will happen if a repository stores highly radioactive materials there? Mathematical models are the

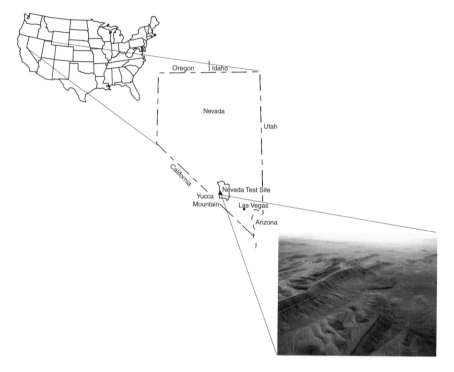

Figure 1.1. Yucca Mountain, in southern Nevada, is the only proposed geological repository site for high-level nuclear waste in the USA. Photograph courtesy of Neil Coleman of US Nuclear Regulatory Commission.

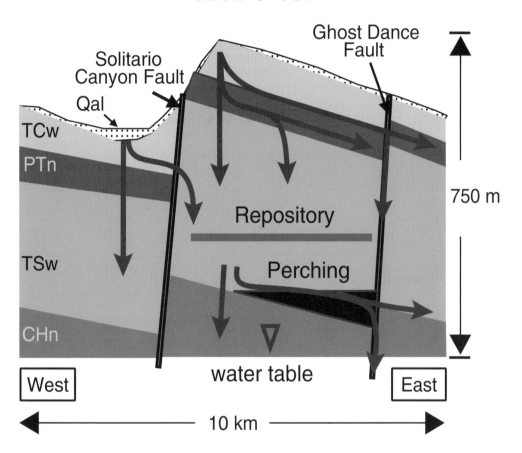

Figure 1.2. The proposed repository is in the thick vadoze zone of volcanic tuff. TCw, PTn, TSw, and CHn stand for tuff units of Tiva Canyon welded, Paintbrush non-welded, Topopah Springs welded, and Calico Hills non-welded. Qal denotes Quaternary alluvium. Courtesy of Jim Winterle at the Center for Nuclear Waste Regulatory Analysis, Southwest Research Institute.

only tools capable of doing that. Secondly, many geochemical processes will affect the safety of the repository. After emplacing high-level wastes, the repository's thermal regime will change and minerals will transform toward achieving a new equilibrium. Radionuclides, if released, will interact with a variety of mineral surfaces, which may retard their movement.

Numerous geochemical modeling studies have been conducted by scientists at the National Laboratories, the Nuclear Regulatory Commission, Southwest Research Institute, and the US Geological Survey. Speciation, solubility, and reaction path models have been used to interpret water–rock interaction experiments (e.g., Knauss and Wolery, 1986), to study radionuclide speciation in groundwater (e.g., Ogard and Kerrisk, 1984) and geochemical evolution (White and Chuma, 1987), and to model surface adsorption reactions (e.g., Turner, 1995; Turner and Pabalan, 1999). Coupled geochemical and transport models are used to evaluate the hydrogelogical systems and reactive transport processes (e.g., Johnson *et al.*, 1998; Winterle and Murphy, 1998). This list only mentions a few of the studies out of the massive number of publications regarding this site.

Geochemical modeling has also been widely used in performance assessment of high-level nuclear waste repositories in other countries, with which we are not as familiar. Therefore, our book has a strong bias toward using examples in the USA.

1.1.2 Mining Related Environmental Issues

Through the whole cycle of a mining project, from the exploration and feasibility studies, permitting, active mining, to remediation, reclamation and closure of a mining site, there is a wide spectrum of environmental issues in which geochemical modeling can play a significant role.

Permitting

In the stage of permitting, i.e., obtaining a permit from state and federal regulatory agencies for mining activities, environmental impact has to be assessed and closure plans have to be approved. Permitting addresses concerns about the events and conditions in the future before mining activities start, and hence the predictive power of geochemical modeling becomes an indispensable tool. Sometimes the results of modeling become contentious and controversial, with the accuracy and validity of geochemical modeling at the center of the storm.

An example of the need for geochemical modeling at this stage of a mining operation is well illustrated by pit lake chemistry and associated risks to wildlife and habitat. Large scale strip mining of gold, particularly in Nevada, will create hundreds of man-made depressions. After the cessation of mining, these craters will fill with groundwater and surface runoff to form *pit lakes* (Figures 1.3 and 1.4). Since these are events that will occur decades in the future, the prediction of pit lake chemistry and risks to aquatic habitat largely depends on geochemical modeling. What else can we use to predict the water quality of a lake that is to be formed 100 years from today?

Figure 1.3. Pit lakes in Nevada, USA. Courtesy of Dr Lisa Shevenell.

Figure 1.4. A pit lake formed from uranium strip mining in Wyoming, USA. Uranium, selenium, and radium are exceeding livestock standards in some pit lakes in Wyoming. Courtesy of Melissa Pratt.

Monitoring

In the active mining phase, environmental monitoring has to be instituted to comply with environmental regulations. The program has to be approved by regulatory agencies, and results have to be submitted to their scrutiny on a regular basis. Geochemical modeling has been used to predict quality deterioration or to diagnose present problems.

Mine Tailings and Mining Waste

Chemical reactions and metal releases are ubiquitous consequences of the extraction, processing, and disposal of geological materials which take place during mining. Typically, milling of ore, the mineralized rock that contains valuable metals, involves grinding the ore into smaller pieces and processing it with various chemical and separation processes. With the advance of new extraction technology, an increasingly smaller amount of metals in the ore (lower grades of ores) can be extracted profitably. That means, for a given amount of metal, an increasing amount of waste is produced. For example, new technology has been developed to extract gold in ores containing only one ounce per ton profitably. Unfortunately, that means that for the amount of gold for a typical wedding ring about three tons of waste materials are produced.

   The barren sandy materials, which remain after the metals have been extracted, are called *tailings*. Tailings often contain toxic metals and chemical agents used in the processing, such as sulfuric acid and cyanide. Tailings are disposed in surface impoundments (tailings ponds) or backfilled in the mining openings.

Acid Mine Drainage

Oxidation of sulfide minerals, occurring either during ore processing or from reactions of tailings and mining wastes with oxygen and water in underground workings, tailings, open pit, and waste rock dumps, produces acidic water rich in toxic metals, commonly referred to as *acid mine drainage* (AMD). The most noticeable environmental changes are the pollution of streams and killing of aquatic life. Sometimes the term "acid rock drainage" is also used, which emphasizes similar occurrences at naturally occurring sulfide ore outcrops.

   Past mining activities have left a legacy of environmental pollution across the world. There are between 100 000 and 500 000 abandoned or inactive mine sites in 26 states of the USA, mostly in the west (King, 1995). The open pit gold mine at Summitville, Colorado, where the lapse of environmental control led to leaks of cyanide-bearing solutions into the Wightman Fork of the Alamosa River, taught a bitter environmental lesson to the nation (King, 1995). In the eastern USA, mining of high sulfur coals in Appalachian states has polluted 11 000 miles (or ca. 17 700 km) of streams and rivers and has endangered drinking water sources. The remediation cost has been estimated at between \$32 to \$72 billion (1 billion $\equiv 10^9$). The US Department of Energy estimates that approximately 10 billion gallons (or 38 million cubic meters) of groundwater are contaminated by uranium mill processing or mill tailings. In Norway, centuries-old mining of copper sulfides has made tens of kilometers of streams and rivers barren of aquatic life. In Canada, there are an estimated 351 million tons of waste rocks, 510

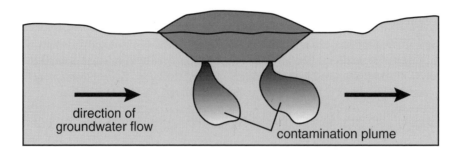

Figure 1.5. A schematic representation of a modern solid waste landfill.

million tons of sulfide tailings, and it is estimated that it will cost about $3 billion to remediate the acid mine drainage problems (Feasby *et al.*, 1991).

1.1.3 Landfills

Landfills are engineered areas where municipal solid waste (commonly referred to as "trash" or "garbage") is placed in certain areas (Figure 1.5). Hazardous wastes have also been subjected to land burial as a means of disposal. The US Environmental Protection Agency (US EPA) estimates that 4.3 pounds (or almost 2 kg) of municipal waste were produced per person per day in 1996 in the USA. This figure is increasing every year. Most of this material finds its way to landfills (Figure 1.6). As a result, there are more than 3 000 active landfills in the USA, and even more inactive landfills (US EPA, 1997). While most new landfills have liner systems and other safeguards to prevent groundwater contamination, some old landfills leak heavy metals and organic pollutants (leachate) to the groundwater and cause public concern. For example, high arsenic concentrations were found in groundwater near the Mt. Trashmore landfill in Virginia Beach, Virginia, in 1998, causing a public outcry.

Speciation, solubility, and surface adsorption modeling are useful tools in simulating the interactions between landfill leachate and liners, the compatibility of mixed wastes, and remediation designs for sites where leachate has already migrated. Potential attenuation of the leachate migration has been the focus of the studies. In particular, geochemical modeling has played an important role in regulatory support for the Hazardous Waste Identification Rule (US EPA, 1995). However, the lack of thermodynamic data for organic–metal complexes has limited the use of geochemical modeling in landfill-related issues.

1.1.4 Deep Well Injection of Hazardous Wastes

Injection of hazardous wastes into deep geological formations has long been a way of waste disposal. Most injection wells are located in the Gulf Coast and the Great Lakes areas in the USA (Figure 1.7). In 1983 alone, 11.5 billion gallons of hazardous waste were injected into deep geological formations in the USA (US EPA, 1985). Waste streams can be very acidic (*p*H 0.03) or alkaline (*p*H 13.8), and can contain toxic organic

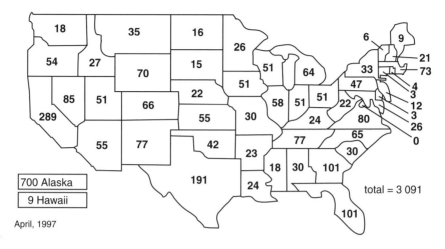

Figure 1.6. The number of landfills in the USA. Adapted from the US EPA web site.

compounds and heavy metals. The injection of waste streams that contrast greatly in chemistry from the host formation can cause severe geochemical reactions (US EPA, 1990). An example is illustrated in Figure 1.8. These reactions have many important implications, such as formation damage, well blowout, and suppression of native microbe communities. Geochemical modeling should have found its most prominent application in this area, although in practice chemical reactions are often simply ignored. This topic is discussed in more detail in §8.5.

1.1.5   Artificial Recharge to Aquifers

Increasing use of groundwater has depleted many aquifers, causing concerns about decreasing availability of groundwater as a resource, land subsidence, and salt water intrusion. One water management technique employed is to replenish depleted aquifers with artificial recharge, or the use of man-made systems to move water from earth surface to underground strata. The US EPA estimates that there are about 1 695 artificial recharge wells in the USA (US EPA, 1999). The recharge water can be surface water (rivers) or treated waste water, and, in most cases, recharging water meets the drinking water or secondary drinking water standards.

However, the mixing of two waters of different chemistry can result in a number of chemical reactions (US EPA, 1999):

- biodegradation by and growth of microorganisms;

- chemical oxidation or reduction;

- sorption and ion-exchange;

- precipitation and dissolution reactions;

- volatilization.

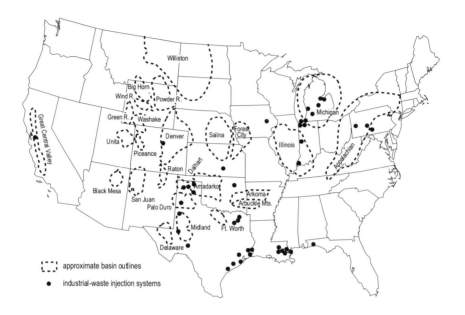

Figure 1.7. Locations of deep-well injection sites and their relation to sedimentary basins in the USA (after Boulding, 1990).

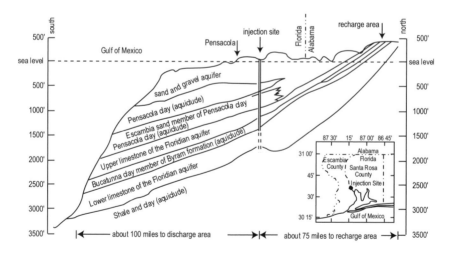

Figure 1.8. An example of a deep-well injection facility. Monsanto's facility near Pensacola, Florida, injects acidic waste streams more than 1000 ft deep into the Lower Limestone of the Floridian Aquifer. Observed reactions include dissolution of limestone, reactions among the wastes under higher pressure and temperature, and suppression of microbial activities (Boulding, 1990).

These chemical reactions may either foul an operation or mobilize harmful substances. Geochemical modeling can be very useful in predicting potential problems, but has only found limited applications.

1.2   The Regulatory Framework

Environmental work is driven by regulatory mandates. It is important to understand the legislation in the country concerned. Here we concentrate on the USA as one example.

1.2.1   CERCLA or Superfund

On December 11, 1980, Congress enacted the Comprehensive Environmental Response, Compensation, and Liability Act (CERCLA) to create the Superfund hazardous substance cleanup program. The Superfund Amendments and Reauthorization Act of 1986 (SARA) made numerous changes to CERCLA to expand the program's scope. The Omnibus Budget Reconciliation Act of 1990 extended the law's taxing authority, which had expired at the end of 1991 under SARA, through December 31, 1995.

CERCLA's purpose is to authorize the federal government to respond swiftly to hazardous substance emergencies and to protect public health and the environment by cleaning up the nation's worst hazardous waste sites (Reisch and Bearden, 1997). The law seeks to make those responsible for the improper disposal of hazardous waste bear the costs and accept responsibility for their actions with a retroactive liability regime, and established the Hazardous Substance Superfund Trust Fund to finance response actions where a liable party cannot be found or is incapable of paying cleanup costs. The Superfund program is the principal federal effort to clean up inactive hazardous waste sites.

1.2.2   RCRA

The Resource Conservation and Recovery Act (RCRA) was enacted in 1976 and amended in 1984. RCRA is focused on hazardous waste treatment, storage, and disposal (TSD) facilities. It provides for "cradle to grave" management and tracking of hazardous wastes, and is designed to prevent the generation of future CERCLA sites. Of special concern in the RCRA regulations is groundwater quality protection. The RCRA regulations have stringent provisions for the design of solid waste and hazardous waste landfills. The RCRA also calls for conservation of energy and natural resources, reduction in wastes generated, and environmentally sound waste management practice.

1.2.3   NEPA

The National Environmental Policy Act (NEPA) of 1969 ensures that all branches of government give proper consideration to the environment prior to undertaking any major federal action that significantly affects the environment. NEPA requires that any project with the potential to harm the environment must have an environmental impact statement

(EIS) and a federal permit. The EIS quantifies the anticipated environmental impact, identifies the benefits, and describes and compares the alternatives.

1.2.4 Clean Water Act

The Clean Water Act (CWA), established in 1972, is the comprehensive national legislation that protects surface waters, including lakes, rivers, aquifers, and coastal areas in the USA. The goal of the CWA is to control toxic water pollutants. The US EPA estimates that, over the last 25 years, the quality of rivers, lakes, and bays has improved dramatically as a result of implementing the public health and pollution control programs established by the Clean Water Act. The total maximum daily load (TMDL) provisions (Sec. 303d) of the CWA are just starting to be implemented. TMDL is defined as the amount of a pollutant that a waterbody can receive and still meet water quality standards. Geochemical modeling will be helpful in assessing the impact of various non-point sources, e.g., natural weathering, groundwater inputs to streams, and fluxes from contaminated stream sediments.

1.2.5 Safe Drinking Water Act

The SDWA was enacted in 1974 to protect the quality of drinking water in the USA. The act authorized the EPA to establish safe health-related drinking water standards, with which all owners and operators of public water systems must comply. The SDWA required the EPA to identify contaminants and rate their potential to harm public health. The wellhead protection provisions of the SDWA are important for groundwater quality protection.

The Underground Injection Control Program (UIC), under SDWA, protects the subsurface emplacement of fluid. This act allowed the SDWA to protect the public from contaminated drinking water. This program is discussed in §8.5.

1.3 The Role of Geochemical Modeling

In §1.1, we described the society's environmental *problems* and *concerns*, and in §1.2, we discussed the regulations that mandate actions to mitigate or solve them. In almost all environmental problems, there is a need for knowledge or predictions of the solute concentrations in space and time. This need can be viewed from two broad categories of field-based environmental problems:

1. contamination problems, and

2. water resource problems.

1.3.1 Contamination Issues

If the concern is contamination or potential contamination of groundwater, the ultimate outcome of a modeler's work is the prediction of solute concentrations in space and time. A typical problem is to determine what will happen to a particular toxic chemical component or species in groundwater downstream from a contamination source, such as a mine tailings pond or polluted industrial manufacturing site. How fast will the contaminant progress downstream, and when will it reach a certain point, such as a property boundary? What processes will slow down its movement (retardation) or immobilize it? Will the concentrations of the contaminant be above regulatory thresholds? Would the remediation methods be effective, i.e., limiting the migration of a contaminant and lowering its concentrations? Alternatively, these questions may be posed for the history of a site concerning past activities, and the question then is not what will happen but what has happened.

Although the general questions outlined above remain the same, answers are sought for different purposes or under different regulatory contexts (e.g., National Research Council, 1990):

- to predict the concentrations in groundwater for risk assessment (exposure). Risk assessment is predicted under current conditions or conditions after remedial actions;

- to evaluate the feasibility of remedial alternatives. (What results will be produced by a given remedial action? How can its performance be optimized?);

- to apportion liabilities among responsible parties (what and/or who caused the contamination, and in what proportions?);

- to apply for a permit for mining or deep well injection or radioactive and hazardous waste storage facilities, etc. Models are used to demonstrate that no potential migration of regulated chemical species or other adverse environmental impacts will be effected or that sufficient measures have been taken to ensure regulatory compliance.

All these questions concern the movement of regulated chemicals in groundwater. Collectively, they are referred to as the "fate and transport problem". Because geochemical reactions affect solute concentrations and their variations in space and time, the need for geochemical modeling is obvious.

1.3.2   Water Resource Issues

A second problem prominent in the environmental arena is the water resource or groundwater management issue, in which geochemical modeling should play a significant role, although it often does not in practice. Here, the water is clean and uncontaminated, but just how much is there to tap and who is entitled to what proportions? In the USA, the problem is most prominent in western and southwestern states where the climate is arid and semiarid, and water resources are limited.

Although water resource issues seemingly concern the movement of water only (flow problems versus transport problems), the chemical constituents in groundwater and their movement may actually help to delineate the flow system that hydraulic data alone fail to reveal. An important limitation for numerical modeling of groundwater flow in the saturated zone is the limited availability of hydraulic parameters and the large uncertainties associated with them (National Research Council, 1990). The geochemistry of a site may provide independent and useful constraints, for example, about recharge areas and rates, flow directions, flow velocities, and inter-aquifer leakage (e.g., Anderman *et al.*, 1996; Keating and Bahr, 1998; Zhu *et al.*, 1998; Zhu, 2000).

Actually, the use of groundwater geochemistry to delineate flow patterns is not limited to clean systems; the same principles apply for contaminated systems. The knowledge of the groundwater flow field at a contaminant site is a prerequisite for development of transport models (National Research Council, 1990). Therefore, contamination problems can also benefit from studies of the movement of dissolved chemicals and isotopes that are not toxic.

From the above discussion, we can see that both "fate and transport" and groundwater resource issues concern the spatial and temporal distribution of solute concentrations. The spatial and temporal distribution of a solute concentration is determined by *mass transport*. Mass transport in the subsurface is a complex process. The concentrations of contaminant or non-contaminant solute in space and time are the outcomes of physical transport processes (advection and dispersion), chemical reactions, and biological activities. The transport equation is used to simulate the movement of solute with the flow of groundwater (advection) and the effects of hydrodynamic dispersion (see Domenico and Schwartz, 1998 and Freeze and Cherry, 1979, for a discussion). For solutes and contaminants that are not reactive or conservative, only physical processes need to be modeled. However, most contaminants and groundwater solutes are reactive. Chemical reactions can retard the migration of a contaminant or transform a chemical into a form with different toxicity. Therefore, geochemical modeling is useful in quantifying these processes that redistribute the mass within or between phases through chemical and biological reactions.

1.4   Current Practice

1.4.1   Model Usage

We hope that the preceding two sections have conveyed a sense of promise and urgency for geochemical modeling. However, statistics show that geochemical modeling has not been used as often as groundwater flow and transport models. We have counted the number of the US EPA's Records Of Decision (RODs) in the 1995 CD-ROM that cited groundwater flow, contaminant transport, and geochemical models. Of the 1 396 total RODs contained in the 1995 CD-ROM, 533 mentioned "groundwater flow model", 93 mentioned "transport model", and 41 mentioned "geochemical model." These statistics (Figure 1.9) illustrate that the applications of geochemical modeling have been limited.

Most applications in the regulatory environment have used speciation–solubility models. A few used surface complexation models. Applications of surface complexation models mostly used the model and data from Dzombak and Morel (1990). Reaction path calculations are mostly limited to the titration and mixing calculations of two fluids.

This limited application is partly due to lack of documentation, and this has prompted us to write this book. In contrast, the environmental applications of groundwater flow

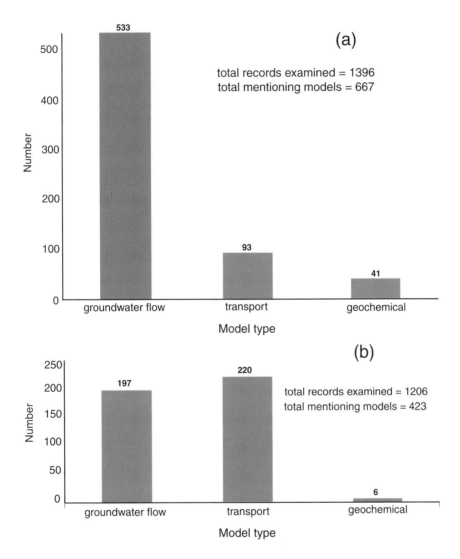

Figure 1.9. Comparison of number of documents that cited the use of groundwater flow, solute transport, and geochemical models. (a) Data from the 1995 Records of Decision (RODs) CD-ROM; (b) data were collected from the September 2000 Superfund Public Information CD-ROM. Source: US EPA.

and solute transport models have been well documented in a number of books, guidance documents, and manuals.

1.4.2 The State of the Art

It must be admitted that perhaps geochemical modeling is not used more extensively because of perceived inadequacies, which result in inaccurate results. These inadequacies are in two main areas.

1. The assumption of (local) equilibrium. It is well known that many surficial processes, whether natural or artificial, are far from chemical equilibrium. Ideally, every such process should be modeled using kinetics, rather than equilibrium thermodynamics. However, our knowledge of such kinetic rates and mechanisms is primitive.

2. Similarly, it is well known that adsorption of solutes on mineral surfaces is extremely important. Our knowledge of such surficial processes in the environmental context is also at an early stage of understanding.[1] [2]

Both areas are subjects of intense and rapidly advancing research at the present time, and it is certain that modeling practice will change considerably in the years ahead. So the question arises as to the usefulness of geochemical modeling at our present state of understanding. To put it another way, is it worthwhile performing the modeling if we know that the results will be inaccurate in some absolute sense?

Actually, we know that *all* modeling of natural processes, such as groundwater flow modeling, is inaccurate to some degree. The question really is whether the model results are useful, and more useful than not modeling at all, or modeling with methods that are known to be less accurate than other methods (such as the almost ubiquitous use of the K_d concept in reactive flow modeling).

We believe that modeling is part of the essence of science; that field or analytical data must be brought together in a conceptual framework, however primitive. As mentioned above, it is also true that modeling *must* be done because of regulatory requirements. Carrying out this modeling always results in increased insight into the problems involved. Therefore, although we will point out inadequacies in geochemical modeling wherever we see them, we nevertheless firmly believe in the value of geochemical modeling in the present environmental context.

1.5 Overview

In the following chapters we attempt to provide an overview of geochemical modeling as practiced today. This includes discussion of the current regulatory framework in the

---

[1] An expanded view of these shortcomings of geochemical modeling is presented in a quotation from Dr John A. Apps on p. 170.

[2] Dr John A. Apps also commented in personal communication that another problem for the under-use of geochemical modeling is the lack of practitioners who understand both geochemistry and hydrogeology. Universities tend to train students specializing in either field but not both.

USA, the various types of models, some of the computer programs commonly used, the databases and how to read them, how to present a problem to the computer program, and how to interpret the results obtained, all illustrated with detailed examples from case histories. In addition, we present a brief introduction to those aspects of the underlying subjects – thermodynamics, surface adsorption, and kinetics – which are necessary to understand the modeling process.

The emphasis is on equilibrium modeling, because this is the dominant practice today. Kinetic modeling will become more and more important in the future. A separate chapter is devoted to kinetic modeling.

Our discussion of geochemical modeling is mostly limited to inorganic contaminants, although a great deal of interest is in the fate of organic chemicals in the environment. The lack of thermodynamic and kinetic properties for organic compounds has limited the use of geochemical modeling in this regard.

2

Model Concepts

2.1 Model Definitions

Voltaire once said "If you wish to discuss with me, define your terms". This seems like a reasonable place to begin; nevertheless, there is always considerable disagreement on the precise definition of complex ideas, and it is often better to accept some imprecision and disagreement, and to get on with the job at hand. We present some brief definitions here simply to clarify the present state of geochemical modeling.

Models

A National Research Council report on groundwater modeling (National Research Council, 1990, p. 52) states that a hydrogeological model comprises three major components:

- specific information describing the system of interest;
- the equations that are solved in the model; and
- the model output.

These are all necessary components for a geochemical model as well, but geochemical models also have one additional component: the equilibrium and kinetic prescriptions for chemical reactions among the chemical components of concern.

In this book we use the term model in a more restricted but more precise way. In our sense:

> A model is an abstract object, described by a set of mathematical expressions (including data of various kinds) thought to represent natural processes in a particular system. The "output data", or the results of the model calculations, generally are quantities which are at least partially observable or experimentally verifiable. In this sense the model is capable of prediction.

As will become clear in later discussions, this definition leaves something to be desired, in the sense that it suggests that a model is *just* a bunch of equations, whereas in fact we make a distinction between computer programs and models. A model is produced by a human being aided by computer programs. He or she must choose what data to use, what computer program options are appropriate, and must select the results which seem reasonable, and reject those which do not.

Because the model uses observational data as input and produces other data which also represent possible observations, it can be said to "represent" or mimic nature. For example, a groundwater flow model uses measured or assumed parameters (e.g., permeabilities and storage coefficients) as input, and produces calculated water velocities, which may or may not be measurable. Models of complex natural situations are thus not only abstractions, but simplified abstractions of nature. In an effort to mimic nature more and more closely, they become more and more complex, often incorporating other models to deal with specific aspects of the overall situation. Thus, for example, a model of water chemistry will probably include a model of activity coefficients (see §3.4.2).

Questions naturally arise as to the accuracy of predictions made by models. Quantum mechanical models of molecular processes are capable of fantastic accuracy. Models of planetary motion based on Newton's laws of motion are sufficiently accurate to put men on the moon and bring them back. Thermodynamics itself is a model of energy relationships, which Einstein once said is the only theory he was sure would never be overthrown. Unfortunately, hydrological and geochemical models deal with much more complex processes, and are less accurate.

2.2   A Holistic View of Geochemical Models

Natural systems, both pristine and anthropogenically perturbed, are complex. Many processes attaining to the movement and distribution of contaminants are coupled or have feedback loops.

Chemical Reactions   For a given amount of mass at a fixed point and time, chemical reactions determine the partitioning of the contaminants among different phases, and the dominant species in the aqueous phase.

Transport   Chemicals are moved or transported by advection – moving water carries the dissolved solids with it – and hydrodynamic dispersion – the spreading and mixing caused in part by molecular diffusion and microscopic variation in velocities within individual pores (Mercer and Faust, 1981).

Biological Processes   Microbial activities can accelerate many chemical reactions that are slow or otherwise kinetically hindered at ambient temperatures. Biological processes can also transform toxic materials into benign compounds.

Fluid Flow   Precipitation and dissolution reactions can increase or decrease the porosity and permeability of the medium and may change the flow velocity. In turn, this can

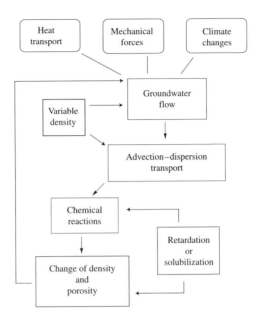

Figure 2.1. Schematic representation of different processes.

affect advective–dispersive transport of the contaminants.

Heat Transport   Fluid flow and chemical reactions are often induced or augmented by the energy flux. Fluids also act as the media for heat conduction.

Figure 2.1 illustrates one of the many possible feedback scenarios. For example, climate changes, mechanical compaction, or heat transport can cause or change groundwater flow. Groundwater flow brings about advection and dispersion of solutes. Mass fluxes across spatial domains cause changes of chemical concentrations, and perturb chemical equilibrium or a chemical steady state, which leads to chemical reactions.

Chemical reactions can cause porosity and permeability changes, which in turn cause changes in groundwater flow. Similarly, chemical reactions can cause changes of density, which in turn will change the flow field.

It is obvious, then, that a geochemical model has to be provided in the context of one part of many processes. The demand for quantitative answers to contaminant concentrations in space and time requires the study of the whole system, not only part of it. The traditional academic exercise of learning *something* about the system is just not enough. At the minimum, for geochemical modeling to be useful, chemical reactions have to be evaluated on a real *time scale* and *spatial coordinates*, which means chemical reactions have to be linked to transport processes.

Because of the complexity of the subjects and the short span of our lives, as well as traditional barriers between different scientific disciplines, different processes are typically studied separately by different specialists. For example, advective–dispersive

---

Total System Performance Assessment – the Yucca Mountain Project

The performance of a geological repository for nuclear waste is influenced by many processes. Determination of whether the total system will comply with regulatory requirements necessitates consideration of all components of the systems and the effects of linking all the components together. This linkage is important because it allows each component to be viewed in the context of the behavior of the entire system. Hence, the concept and methodology of Total System Performance Assessment (TSPA) is widely used in nuclear waste disposal (Figure 2.2).

The US Department of Energy's TSPA report for the Yucca Mountain Site Characterization Project (TRW, 1999) gave the following description of linkage between the processes:

> For example, waste package material degradation may be characterized by laboratory tests of corrosion. However, the geologic system in which the waste is to be emplaced is analyzed using field studies of the host rock for properties that are only observed on a large scale (such as fracture density), as well as laboratory studies of other aspects (such as water chemistry). In a functioning system, these elements provide feedback to one another. The interaction of the water with the corroded waste would likely change the water chemistry, which may in turn change the fractures and the way water flows through them. This very simple example shows an obvious potential for feedback. When a very complex system with numerous components is simulated as a single system in a TSPA, interactions among the components that would not otherwise be observed are often found in the analysis.

---

mass transport is modeled by hydrogeologists; and chemical reactions are modeled by geochemists.

Because of the demand for solving environmental problems and the explosion of computing power at ever reduced prices, there has been tremendous growth in models of coupled processes. Nevertheless, the complexity of nature and demands on computing power still preclude models that include all the processes discussed above. Very often we are forced to consider simplified subsets of the general problem (Mercer and Faust, 1981).

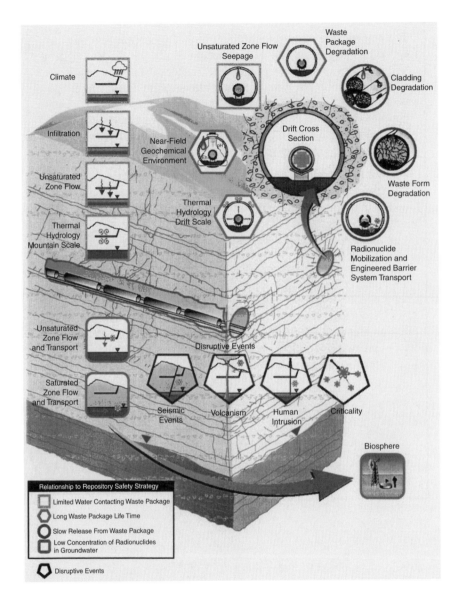

Figure 2.2. Schematic representation of the concept of total system performance assessment at Yucca Mountain, Nevada, USA. Source: US Department of Energy Yucca Mountain Project web site (www.ymp.gov).

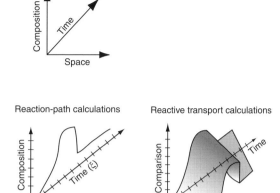

Figure 2.3. Different levels of complexity of geochemical models. ξ refers to reaction progress variable (see below). After Raffensperger (1996).

2.3   Types of Geochemical Models

Geochemical models have been extensively reviewed in the literature (Appelo and Postma, 1993; Nordstrom and Ball, 1984; Paschke and van der Heijde, 1996; and Plummer *et al.*, 1992). Here we describe them only briefly. More detailed descriptions of models can be found in the above-mentioned reviews, the manuals for the geochemical modeling codes, and individual chapters for respective types of models in this book.

   In general, geochemical models can be divided according to their levels of complexity (Figure 2.3). Speciation–solubility models contain no spatial or temporal information and are sometimes called zero-dimension models. Reaction path models simulate the successive reaction steps of a system in response to the mass or energy flux. Some temporal information is included in terms of reaction progress, ξ, but no spatial information is contained. Coupled reactive mass transport models contain both temporal and spatial information about chemical reactions, a complexity that is desired for environmental applications, but these models are complex and expensive to use.

2.3.1   Speciation–solubility Models

Speciation–solubility geochemical models can answer three questions, given concentrations of constituents measured analytically in a system at the temperature and pressure of interests:

1. What are the concentrations and activities of ionic and molecular species in an aqueous solution?

2. What are the saturation states with respect to various minerals in the system, and hence the directions of reactions that might occur toward achieving equilibrium?

3. What is the stable species distribution on surfaces or ion-exchangers that is at equilibrium with the aqueous solution?

Speciation–solubility models deal with a closed, static, batch or beaker-type system. However, speciation–solubility models also serve as the basis for the reaction path and reactive transport models discussed below. Equilibrium calculations are also useful to evaluate kinetic rates as a function of the deviation from equilibrium. Direct applications of speciation–solubility models include assessment of bioavailability because the toxicity of some contaminants (e.g., chromium and arsenic) varies drastically for different species.

Two types of algorithm are used in speciation–solubility models: those that use equilibrium constants and those that use free energy minimization. Most geochemical models we described in the text belong to the first type. Interested readers are referred to Anderson and Crerar (1993) for a more detailed discussion.

2.3.2 Reaction Path Models

Reaction path models calculate a sequence of equilibrium states involving incremental or step-wise mass transfer between the phases within a system, or incremental addition or subtraction of a reactant from the system, possibly accompanied by an increase or decrease of temperature and pressure (Helgeson, 1968, 1969). The calculated mass transfer is based on the principles of mass balance and thermodynamic equilibrium. Unlike speciation–solubility calculations, which deal with the equilibrium state of a system, the reaction path model simulates processes, in which the masses of the phases play a role.

Although all reaction path models are based on the principle of mass balance and thermodynamic equilibrium, various configurations of the mass transfer models have been constructed to simulate different processes.

Titration Mixing or titration modeling simulates the process of addition of a reactant into a system (Figure 2.4). The reactant can be a mineral, a chemical reagent, a glass, a gas, another aqueous solution, a "rock", or anything for which the chemical stoichiometry can be defined. Evaporation is considered as a negative titration of water. In a titration or mixing model, after each aliquot of the reactant, the aqueous solution is re-equilibrated by precipitating or dissolving a solid or gas phase or phases. The mixing and titration models have been most widely used in the environmental field because they are the most straightforward, and many environmental processes resemble a titration or mixing process. Titration models are discussed in more detail in §8.1.

Buffering A special configuration of mass transfer modeling is where the activity of a gas species or an aqueous species is fixed. Environmental systems or processes resembling this configuration include laboratory experiments in which the pH is buffered or reactions at the Earth's surface in which the atmosphere acts as a large external gas reservoir. Bethke (1996) gives an example of pyrite oxidation where the fugacity of oxygen gas is fixed by atmosphere and compares it with the situation where oxygen supply is limited.

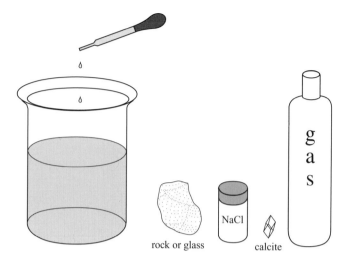

Figure 2.4. Schematic representation of a mixing or titration model. The reactant can be a mineral, a chemical reagent, a glass, a gas, another aqueous solution, a "rock", or anything for which the chemical stoichioimetry can be defined.

Flush The "flush" reaction path model is analogous to the "perfectly mixed-flow reactor" or the "continuously stirred tank reactor" in chemical engineering (Figure 2.5). Conceptually, the model tracks the chemical evolution of a solid mass through which fresh, unreacted fluid passes through incrementally. In a "flush" model, the initial conditions include a set of minerals and a fluid that is at equilibrium with the minerals. At each step of reaction progress, an increment of unreacted fluid is added into the system. An equal amount of water mass and the solutes it contains is displaced out of the system. Environmental applications of the "flush" model can be found in simulations of sequential batch tests. In the experiments, a volume of rock reacts each time with a packet of fresh, unreacted fluids. Additionally, this type of model can also be used to simulate "mineral carbonation" experiments.

Kinetic Reaction Path Model Although the original reaction path models are built on the concept of "local equilibrium" and "partial equilibrium", and our discussions above all concern equilibrium models, kinetics also can be incorporated into reaction path models (Gunter *et al.*, 2000; Lasaga, 1998). In these models, homogeneous reactions in the aqueous phase are assumed to be controlled by equilibrium. Heterogeneous reactions, i.e., dissolution and precipitation of solids, are controlled by kinetics. Very different results are obtained when kinetics is involved in the classical feldspar dissolution example (Gunter *et al.*, 2000; Lasaga, 1998).

It should be noted that the "flush" model, other reaction path models, such as the "fluid-centered reaction path" model, and models with the "dump" option (see Wolery, 1992), have become less useful for their originally intended uses in simulating reactive transport. Although the extent of reactions is often monitored by the reaction progress variable (ξ), no temporal information is included in the model. Additionally,

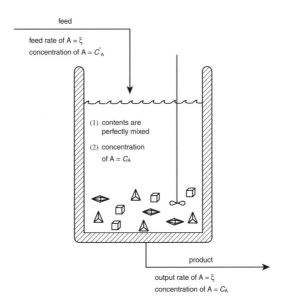

Figure 2.5. Schematic representation of the "perfectly mixed-flow reactor" model. A refers to the component of interest; C denotes the concentration; and ξ stands for the feed rate or reaction progress variable.

no spatial compositional change is included in the models (Lichtner, 1996; Steefel and MacQuarri, 1996). Recent development of coupled reactive transport models can deal with reactions in real spatial and time scales as well as physical transport processes such as advection and hydrodynamic dispersion (see Chapter 10). Thus these reaction path models' applications to environmental problems lie in the special situations discussed above.

2.3.3   Inverse Mass Balance Models

The mass balance concept in the inverse mass balance models is quite simple (Figure 2.6). If one fluid is evolved from another, the compositional differences can be accounted for by the minerals and gases that leave or enter that packet of water:

$$\text{initial water} + \text{reactants} \rightarrow \text{final water} + \text{products}$$

Another possibility is that the final water may evolve from the mixing of two initial waters, and the mixing fractions then become part of the mass balance calculations.

In hydrogeology, because it is the reactions that have occurred along a flow path that are to be modeled, inverse mass balance modeling is often called *reaction path modeling*. However, "reaction path" here means something quite different from that defined by Helgeson (1969) and widely used in high-temperature geochemistry (see §2.3.2).

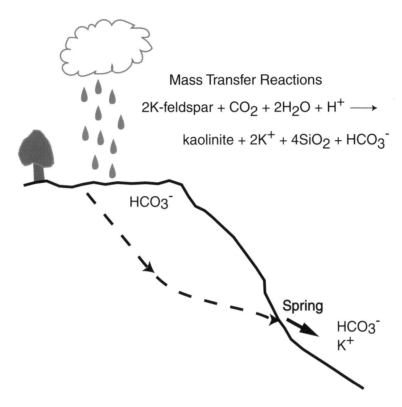

Mass Transfer Reactions

$$2\text{K-feldspar} + CO_2 + 2H_2O + H^+ \longrightarrow$$

$$\text{kaolinite} + 2K^+ + 4SiO_2 + HCO_3^-$$

HCO_3^-

Spring

HCO_3^-
K^+

Figure 2.6. Schematic representation of the inverse mass balance model. If we know that the spring at the foothill is evolved from rain water, possible mass transfer reactions can be modeled from the mass balance principle.

Inverse mass balance modeling here only employs the mass balance principle; thermodynamics and equilibrium are not considered. Inverse models are usually non-unique. A number of combinations of mass transfer reactions can produce the same observed concentration changes along the flow path. *Mass transfer reactions* here refer to the reactions that result in the mass transfer between two or more phases, such as the dissolution of solid and gas or precipitation of solids. Chapter 9 describes the details of the models and shows a few examples.

2.3.4   Coupled Mass Transport Models

As discussed above, the processes that affect the partitioning of contaminants between phases and the movement of the contaminant are coupled in nature. These processes affect each other, and the ultimate fate and transport of contaminants is the cumulative effect of all these processes.

In this book, we use the term *coupled model* to describe models in which two sets of equations that describe two types of processes are solved together. For example, *multi-component, multi-species coupled reactive mass transport models* (this is a mouthful,

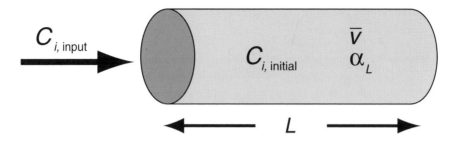

Figure 2.7. Schematic diagram of a one-dimensional reactive transport model. Both transport processes (advection–dispersion) and chemical reactions are included in the model. C stands for concentration, $\bar{v}$ for groundwater velocity, L for length, and α_L for longitudinal dispersivity.

see Figure 2.7) refer to those models that include solutions for both the partial differential equations that describe advective–dispersive transport and solutions for the algebraic equations that describe chemical equilibrium. In these models, chemical reactions are evaluated on the temporal and spatial scales.

By this definition, *reactive transport models* that use an isotherm to describe the partitioning of a contaminant between groundwater and the solid matrix are not coupled models. In these models, only one type of equation, the advective–dispersive–reactive equation, is solved.

Coupled mass transport models can also include heat transport (e.g., Raffensperger and Garven, 1995) or fluid flow. Coupled reactive transport models represent the desired tools for evaluating fate and transport of contaminants.

2.4   Model Verification and Validation

The usefulness of geochemical or other complex models would seem to be determined in two stages.

1. Model *verification*. Does the model do what it is intended to do? That is, does the computer code solve the equation set correctly, and is the code free from serious bugs that will lead to unintended results?

2. Model *validation*. Does the model produce realistic results, given realistic input data? That is, are the underlying conceptual models which the coded equations simulate adequate? Do they actually "represent" natural processes?

It would seem to follow, then, that after verification and validation, a model could be used for predictive purposes. If only things were that simple. A useful discussion of these concepts is given by Nordstrom (1992). We present only a synopsis.

Theoretically, verification of a computer code is an achievable task, given sufficient time, funding, and manpower. However, with increasingly complex programs, complete and unambiguous verification is becoming very difficult, or even infeasible. Nordstrom

(1992) presents a number of suggestions that should be adhered to in verifying computer codes. Normally, however, this is not a problem for the user of the codes (for whom the present text is written), because codes which are generally available and widely used have all been verified. They do what they are supposed to do.

Validation, the process of determining whether the program simulates nature in some useful way, is conceptually a much trickier issue. Konikow and Bredehoeft (1992), Nordstrom (1992), and Oreskes *et al.* (1994) argue convincingly that in fact

> . . . geochemical models cannot be validated by field data.

Bredehoeft and Konikow (1993) state

> . . . using the terms validation and verification are misleading at best. These terms should be abandoned by the groundwater community.

The problem can perhaps be summarized by considering the point of view of a non-scientist working for some regulatory agency, who has a report on the results of some geochemical modeling which predicts the concentrations and dimensions of some contaminated waters 100 years in the future. He or she reasons that if the programs used in the modeling have been *validated*, then the results are true in an absolute sense. But no scientist would believe this, no matter how good the models, in our present state of knowledge.

We subscribe to the view held by many active workers in the field, exemplified by the following quotation from Walter *et al.* (1994):

> With increasing model complexity and multiple interacting processes, complete and unambiguous verification and validation may be impossible due to the unavoidable non-uniqueness of simulation results. However, the lack of general validation does not detract from the value of the model, which is, above all, a tool to give insight into complex processes.

> We will therefore avoid the use of the terms "verification" and "validation" and use instead "model evaluation" and "model comparison", [1] which we believe reflect more accurately what actually takes place and leave us with a realistic sense of the model's (and our own) limitations. These limitations cause no problem if one is aware of them and if the model is used primarily to gain insight into the interactions of the various physical and chemical processes included in the model.

Of course, the models are not only used "to gain insight", but to make predictions required by regulatory bodies, and to evaluate proposed remedial procedures which may be very expensive. As mentioned, these predictions may well be inaccurate, but they must be made.

The interaction of waste with geological media involves complex and interacting processes. One important use of mathematical models is the integration of data, concepts, and processes that are from different disciplines. There is no better way to understand them than to construct geochemical and hydrological models, and the art of

---

[1] We'd disagree with the term "model comparison", which to us means code comparison.

constructing such models will improve with time and experience. It may be disappointing to new users of geochemical models that we must always distrust our results, for reasons that will be discussed in more detail in later chapters, but that is the case, and it is also true that modeling is at the same time one of the surest methods to improved understanding.

2.5 Model Usefulness and Limitations

In Chapter 1, we presented the value of understanding spatial and temporal distributions of solute concentrations in environmental practice. Geochemical modeling can be extremely useful if the modeling results can contribute to this goal. This means that chemical reactions have to be modeled on a real *time scale* and with real *spatial coordinates*. In general, this goal is only partially achievable at present using geochemical modeling.

Accurate model prediction, no matter how practical it is, can be obtained only by understanding the underlying science. In this case, the prediction of mass transport in the subsurface can only be accurate if we understand the geochemical reactions that occur in the aquifer. This requires that we understand the geochemical properties of the aquifer, have the thermodynamic and kinetic properties of the chemical system in hand, and understand the interplay among chemical, physical, and biological processes.

Currently, there are a number of fundamental processes in geochemistry that are not well understood, and which limit the applications and usefulness of geochemical modeling.

1. Most models only simulate the batch type of chemical systems, and lack the time and spatial information needed to address a site-specific problem, e.g., the arrival of a contaminant at a particular point in space. This type of model, however, can be useful in evaluating some laboratory experiments.

2. There is a general lack of kinetic data for critical environmental and geochemical processes. Additionally, despite recent advances in chemical kinetics, the applicability of laboratory kinetic data to field situations is still uncertain. There is a large discrepancy between laboratory and field derived rate constants (Lasaga, 1998; Zhu *et al.*, 2001b).

3. Although the importance of surface adsorption is well recognized, critical issues still remain such as how to characterize the effective surface areas of the sorbents and the applicability of laboratory data to the field.

4. The lack of equilibrium among redox couples in natural aquatic systems is well known, but how to handle it in the models is unclear.

5. Chemical heterogeneity in the solid matrix of aquifers is usually not well characterized or appreciated. However, chemical heterogeneity can play an important role in the assessment of model uncertainty (e.g., Zhu and Burden, 2001).

6. Modeling results bear social and economic consequences (consider, for example, the question "is the nuclear waste repository safe?") There is a need for qualitative or quantitative evaluation of modeling uncertainties. However, uncertainties in modeling results are seldom provided.

Because of this current state of knowledge, a large part of this book will focus on geochemical equilibrium models. Chemical equilibrium models are the best that we can do for most geochemical processes. We may argue that understanding equilibrium is the good starting point for understanding kinetics.

Although it can be said that "if you cannot model a natural system, you don't understand it", it must also be said that "even if you can model it, you may still not understand it completely". Given the complexity of many environmental problems, it is almost certain that we " . . . will not understand it completely". It follows that any predictions we make based on our model will be inaccurate to some degree. This is certainly true for the contaminant transport models that are currently used to predict concentration distribution in space and time (see discussion in §10.3).

In other words, geochemical (and hydrological) models of natural and engineered systems at the present time are, at best, approximations to the real situation. To meet various regulatory requirements, these approximations, and predictions based on these approximations, must be made, and there is no better way to proceed than to construct models based on fundamental science. To this extent, geochemical and hydrological modeling is useful.

It is well known that model predictions made in the past have often turned out to be highly inaccurate. However, we learn from every new case, and there have been many success stories too. It is an article of faith among environmental modelers, just as among scientists in any field, that with increased knowledge will come an increase in our ability to understand and to model our environment.

The above discussion of the difficulties appears to point out a "mission impossible". We probably never will fully understand everything, at least not any time soon. On the other hand, the challenges also present opportunities for geochemists at two different levels. On the scientific level, the studies of geochemical reactions in active systems where the boundary conditions can be better defined, samples are readily available, and modeling results can be verified in short periods of time, may help to change the scientific paradigm in earth sciences. We also now have the funding and urge to tackle problems such as kinetics, redox reactions, and surface adsorption.

On a second level, opportunities exist because the current level of geochemical studies in the environmental field is low. We know that a distribution constant K_d is usually insufficient to describe the chemical processes (see §10.3), but most "fate and transport" models have employed this concept. Geochemical modeling is generally under-used.

In the examples that follow, we will attempt to address each of these components. The method we have adopted is to augment the narrative with "Boxes" containing additional specialized information or advice, explanation of tricky points, and the relation of the material to fundamental thermodynamics.

3

Thermodynamic Background

Geochemical modeling programs are built, for the most part, on the fundamental laws of thermodynamics and kinetics. In this chapter we introduce the subject of equilibrium thermodynamics. Kinetics is dealt with in Chapter 11.

Although it is not necessary to be able to derive all the relationships used in modeling from first principles, it is necessary to have a reasonable understanding of the meanings of the terms used. These are, of course, derived and discussed in innumerable textbooks on the subject (e.g., Anderson, 1996; Anderson and Crerar, 1993), but they can be difficult to find in the large amount of material covered, or difficult to understand in the context of natural or environmental systems. An attempt is made here to present the fundamental principles in a reasonably intuitive way, omitting some tedious derivations. We begin by presenting definitions of the essential concepts.

3.1   Systems and Equilibrium

3.1.1   Real and Model Systems

In a complete treatment of thermodynamics, systems are divided into various types, the details of which we can safely neglect here. But there is one point about systems that we should not neglect, and that is that thermodynamic systems and real systems are two different things.

The aim in performing geochemical modeling (or in fact most kinds of modeling) is to learn something about some natural system. In environmental problems, the "natural system" is normally at some geographic location, has dimensions of some hundreds of meters horizontally and perhaps some tens of meters vertically, and is made up of rocks, soils, and fluids in various combinations. We want to understand what processes are occurring in the system, and to predict the future state of the system. The exact definition of what constitutes this natural system – its dimensions, and what is in it – is arbitrary, but is usually fairly obvious, given the nature of the problems involved. These real or natural systems are almost always changing, slowly or rapidly: fluids are flowing, and chemical reactions are occurring between fluids, rocks, and soils.

However, when we perform thermodynamic calculations in order to better understand this natural or geological system, strictly speaking our calculations refer not to this real system, but to a thermodynamic model of the real system. Naturally, we want the thermodynamic model to mimic or represent the real system closely, because otherwise the results of the model will be useless. Nevertheless, even if the thermodynamic system is as good and as faithful as we can make it, there are two important differences.

The Differences Between Real and Model Systems

First, the model or thermodynamic system will virtually always be incomplete and inaccurate to some degree. The reasons for this have largely to do with the choice and quality of the data used by the model. This will be discussed in more detail in later sections (e.g., §4.1). The goal of modeling could be said to be to minimize this degree of incompleteness and inaccuracy.

Secondly, models based on thermodynamics virtually always refer to a state or states of complete equilibrium. How can these models be useful in understanding real systems, which, as we have noted, are constantly changing? It is an interesting paradox.

3.1.2   Equilibrium

We begin with the fairly intuitive idea that all systems, whether mechanical or chemical, will, left to themselves, spontaneously lower their energy content to the lowest possible level. In other words, weights fall until they can fall no further, and reactions proceed until they reach equilibrium. Mechanical systems spontaneously lower their *potential energy*, and chemical systems spontaneously lower another kind of energy (*Gibbs energy*) to be discussed later in this chapter.

A system is said to be at *equilibrium* if none of its properties change with time (obviously eliminating virtually all environmental systems at one stroke). This definition includes *metastable equilibrium* states, which are not at their lowest possible energy level, but are *constrained* from changing to a lower energy level; and *stable equilibrium* states, which have the lowest energy level available, and cannot change to any lower energy state. Thus, diamond is a metastable form of carbon under ordinary conditions, and graphite is the stable form.

Thermodynamics allows us to calculate the energy differences between equilibrium states of all kinds, stable and metastable, for all kinds of substances. If that is all it does, and natural systems are not at equilibrium, how can it be useful?

Local Equilibrium

Even though a natural system may not be at equilibrium overall, there may well be small parts of the system which *are* at or not very far from equilibrium. The system is then said to have areas of *local equilibrium*, and thermodynamics can be applied to these smaller parts. For example, a solution flowing through a limestone may be in the process of dissolving calcite – the system is not at equilibrium. However, the calcium and carbonate and other ions in the solution may well be at or not far from equilibrium among themselves. In other words, a portion of the solution, if removed from contact with the limestone, would not change – it is in a state of local equilibrium, and thermodynamics can be used to show that the solution is in fact undersaturated with calcite, and will dissolve calcite if in contact with it.

Other systems may have gradients in temperature or composition, meaning that they are not at equilibrium. Again, small parts of the overall system may be chosen such that the gradient in that small part is negligible, and thermodynamics can be applied to that small area of local equilibrium.

Consider a combination of processes, such as the solution mentioned above dissolving limestone in one area, then flowing to another location where it loses some of its CO_2 content, and precipitates calcite. The overall process is, of course, far from equilibrium. Nevertheless, the process can be considered in a number of separate steps, each of which is not far from equilibrium. Thus a state of local undersaturation might be calculated, then the calcium and carbonate content of the solution increased slightly, or the CO_2 content decreased slightly, or whatever, and the calculation repeated until some final state is achieved. The overall process is simulated as a number of equilibrium steps. This is a type of geochemical modeling known as *titration*, and is discussed in Chapter 8.

In other words, because thermodynamics only applies to equilibrium states, our geochemical models apply only to areas of local equilibrium, and therefore we can only successfully model natural systems which have areas of local equilibrium. But it is in fact very difficult to determine whether natural systems do have such areas of local equilibrium, and on what scale. This problem will be discussed in more detail at the end of this chapter (§3.11).

3.1.3 The Role of Kinetics

Because thermodynamics deals only with equilibrium states, we can use it to show, for example, that calcite should dissolve in a certain solution, or that it should precipitate from another solution, but we are unable to say anything whatsoever about *how fast* such processes will occur, or indeed if they will occur at all (metastable supersaturated solutions are well known, after all). This is a serious limitation. Many important processes may be *rate-limited* by one or more slow reactions.

Kinetics is the science which deals with the mechanisms and rates of chemical reactions, and ideally kinetic models should be incorporated into geochemical models, along with thermodynamics. This is being done increasingly, and is the subject of Chapter 11. The rest of this chapter outlines those aspects of thermodynamics needed to understand geochemical models.

3.2 Chemical Reactions

Pretty well everything we will be doing boils down to the determination of what chemical reactions are important, and determining whether that reaction is proceeding to the right, or to the left, or is at equilibrium. For example, we often want to know whether a particular mineral is dissolving or precipitating. We write

$$\text{solid mineral} \rightleftharpoons \text{dissolved mineral} \tag{3.1}$$

If this reaction proceeds to the right, the mineral is dissolving. If it proceeds to the left, it is precipitating. In most cases, the dissolved form of the mineral is ionized, so we write, say, for calcite,

$$CaCO_3(s) = Ca^{2+} + CO_3^{2-} \tag{3.2}$$

where (s) means a solid phase (and (g) means a gas phase, and (aq) an aqueous solute, although we don't bother adding (aq) for ions). We usually use the $=$ sign rather than $\rightleftharpoons$ to be completely general.

Some minerals do not form ions on dissolving, at least under normal conditions. For example, for quartz we write

$$SiO_2(s) = SiO_2(aq) \tag{3.3}$$

where $SiO_2(aq)$ is the dissolved form of silica.

Similarly, for a gas we could write

$$CO_2(aq) = CO_2(g) \tag{3.4}$$

to determine whether CO_2 will dissolve into the aqueous phase or exsolve out of it.

Having written the reaction of interest, we now need to know which way it will go under our chosen conditions. This is done by determining the energy per mole of each product and reactant. If the products have more energy than the reactants, the reaction goes to the left, and vice versa. However, a special kind of energy is required.

3.3 Gibbs Energy

It can be shown that, because we always consider reactions at a given temperature (T) and pressure (P), the appropriate energy is the Gibbs energy, G. So for (3.2), if

$$G_{Ca^{2+}} + G_{CO_3^{2-}} > G_{CaCO_3(s)} \tag{3.5}$$

calcite will precipitate, and vice versa. Because the magnitudes of $G_{Ca^{2+}}$ and $G_{CO_3^{2-}}$ depend on concentration, precipitation lowers both quantities, and will proceed until

$$G_{Ca^{2+}} + G_{CO_3^{2-}} = G_{CaCO_3(s)} \tag{3.6}$$

at which point the reaction will be balanced, or at *equilibrium*.

Unfortunately, it is not possible to measure values of G of any substance. Only *differences* in G are measurable. Therefore, for every substance of interest, solid, liquid, gas, or solute, we measure, usually by calorimetric methods, the quantity $\Delta_f G°$, which is the difference between G of a compound substance, and the sum of the G values of its constituent elements, each in its most stable state. For example, for calcite,

$$\Delta_f G° = G°_{CaCO_3(s)} - G°_{Ca} - G°_{C} - 1.5\,G°_{O_2(g)} \tag{3.7}$$

and this quantity is tabulated in databases for solids, liquids, gases, and solutes. Of course, as this quantity varies with temperature and pressure, and for solutes with concentration as well, it must be tabulated for one specified set of conditions, called the *standard state* of the substance, and this fact is noted by the superscript $^\circ$. It is measured in joules per mole ($J\,mol^{-1}$) or calories per mole ($cal\,mol^{-1}$), where $1\,cal = 4.184\,J$.

Therefore, if all products and reactants in a chemical reaction are in their respective standard states, we can tell which way the reaction will go. For reaction (3.2), for example, we look up (or our computer program looks up) the values of $\Delta_f G^\circ$ for each product and reactant, and we calculate

$$\Delta_r G^\circ = \Delta_f G^\circ_{CaCO_3(s)} - \Delta_f G^\circ_{Ca^{2+}} - \Delta_f G^\circ_{CO_3^{2-}} \tag{3.8}$$

where $\Delta_r G^\circ$ is the *standard Gibbs energy of reaction*, following the convention that Δ_r means *products−reactants*. As mentioned, if this quantity is negative, the reaction proceeds as written, and calcite precipitates. If it is positive, calcite dissolves, all assuming standard state conditions (the conditions for which we have tabulated data). The fact that we have a bunch of G terms for the elements in there does not matter, because they all cancel out in balanced reactions, and our reactions are of course always balanced.

3.3.1 Enthalpy and Entropy

In introducing the Gibbs energy directly, we have short-circuited a great deal of the usual methodical derivation. It will be sufficient for our purposes to say simply that the formal definition of G is

$$G = H - TS \tag{3.9}$$

or, for a chemical reaction,

$$\Delta_r G = \Delta_r H - T\Delta_r S \tag{3.10}$$

or, with standard state values,

$$\Delta_r G^\circ = \Delta_r H^\circ - T\Delta_r S^\circ \tag{3.11}$$

where H is the molar *enthalpy*, $\Delta_r H$ is the *heat of reaction*, the amount of heat absorbed or given off during the reaction as written, and $\Delta_r H^\circ$ is the *standard* heat of reaction, when all products and reactants are in their standard reference states.

S is the molar *entropy*, a measure of the disorder in the system, and $\Delta_r S$ and $\Delta_r S^\circ$ are the entropy changes in a reaction, using either the real entropies, or the standard state entropies. All these Δ terms use the *products−reactants* convention.

Finally, although neither G nor H can be measured in absolute terms, so that we are forced always to use differences of these quantities, absolute values of S *can* be measured calorimetrically. Thus, tables of thermodynamic data for compound i contain values of $\Delta_f G^\circ_i$, $\Delta_f H^\circ_i$, and S°_i, where S°_i is the entropy of i. If we want a value of $\Delta_f S^\circ_i$, we must calculate it from the tabulated values of S°_i for the compound and its constituent elements.

A typical table of thermodynamic data looks like Table 3.1. Note that $\Delta_f H^\circ$ and $\Delta_f G^\circ$ for the element Ca are both zero. This does not mean that $H^\circ_{Ca} = 0$, or $G^\circ_{Ca} = 0$. We don't know what these values are. $\Delta_f G^\circ = 0$ just means that $G^\circ_{Ca} - G^\circ_{Ca} = 0$.

Formula	Form	Mol. wt (g mol^{-1})	$\Delta_f H°$	$\Delta_f G°$ (kJ mol^{-1})	$S°$	$C_p°$ (J mol^{-1} K^{-1})	$V°$ (cm^3 mol^{-1})
Calcium							
Ca	s	40.0800	0	0	41.42	25.31	
Ca^{2+}	aq	40.0800	-542.83	-553.58	-53.1	—	-18.4
CaCO$_3$	calcite	100.0894	-1206.92	-1128.79	92.9	81.88	36.934
CaCO$_3$	aragonite	100.0894	-1207.13	-1127.75	88.7	81.25	34.150
Carbon							
C	graphite	12.0112	0	0	5.740	8.527	5.298
C	diamond	12.0112	1.895	2.900	2.377	6.113	3.417
CO$_3^{2-}$	aq	60.0094	-677.149	-527.81	-56.9	—	-6.1
HCO$_3^-$	aq	61.0174	-691.99	-586.77	91.2	—	24.2

Table 3.1. A typical table of thermodynamic data (NBS tables; Wagman *et al.*, 1982).

3.4 Activity, Fugacity, and Chemical Potential

3.4.1 Activity and Fugacity

We rarely have occasion to consider reactions where all products and reactants are in their standard states. Therefore, we need another difference term, the difference between $G°$ of each product and reactant in its standard state, and G of each in the real state we are interested in. This is the function of the *activity*, a (dimensionless) quantity which tells us this difference. Thus, for any substance (solid, liquid, gas, solute, or ion) i, we define the activity a such that

$$G_i - G_i° = RT \ln a_i \qquad (3.12)$$

where G is the Gibbs energy per mole of i in our system, $G_i°$ is the Gibbs energy per mole of i in the standard state, and R is the *gas constant* (8.31451 J K^{-1} mol^{-1}, or 1.98722 cal K^{-1} mol^{-1}).

Standard states have been chosen such that if our system behaves ideally (obey's some simple rules such as the ideal gas law for gases; Henry's law, Raoult's law for solutes), the activity takes on very simple forms, and, if the system is not ideal, we introduce a fudge factor called the *activity coefficient* to convert the very simple form into the true activity.

For solid and liquid solutions:

$$a_i = X_i \gamma_{R_i} \qquad (3.13)$$

For gaseous solutions:

$$a_i = P_i \gamma_{f_i} \qquad (3.14)$$
$$= f_i \qquad (3.15)$$

For aqueous solutions:

$$a_i = m_i \gamma_{H_i} \qquad (3.16)$$

where a_i is the activity of any substance i, X_i is its mole fraction, m_i is its molality, P_i is its partial pressure (total pressure × mole fraction), and f_i is called the *fugacity*. The fudge factors are γ_{R_i}, γ_{H_i}, and γ_{f_i}, and are measures of the deviation from the ideal behavior of substance i from Raoult's law, Henry's law and the ideal gas law, respectively.

Note that, forgetting about activity coefficients for the moment, we have arranged things rather conveniently, such that the activity of pure solids and liquids will be 1.0 (the mole fraction of a pure compound being 1.0); the activity of a solid solution component is its mole fraction; the activity of a gas is (numerically equal to) its partial pressure; and the activity of an aqueous solute is (numerically equal to) its molality. These are useful approximations for real systems, which can be improved by using the activity coefficients. Note that activities are always dimensionless, though we have not demonstrated this.

3.4.2 Activity Coefficients

Aqueous Ionic Species – Henryan Coefficients

The Debye–Hückel Equation   Activity coefficients (γ_H) for ions can be calculated for relatively low concentrations by variations of the Debye–Hückel equation. The "extended" Debye–Hückel equation is

$$\log \gamma_{H_i} = -Az_i^2 \frac{\sqrt{I}}{1 + B\mathring{a}\sqrt{I}} \qquad (3.17)$$

where A and B are temperature-dependent constants and $\mathring{a}$ is an adjustable parameter corresponding to the size of the ion. The ionic strength I is defined as

$$I = \frac{1}{2} \sum_i m_i z_i^2 \qquad (3.18)$$

where m_i is the molality of ionic species i, and z_i is its charge.

To calculate the activity coefficient of ions, virtually all geochemical modeling programs today use either a variation of the Debye–Hückel equation or the Pitzer equations. Two variations of the Debye–Hückel equation in common use are the Davies equation and the B-dot equation.

Note that for NaCl the ionic strength is almost the same as the molality, because the dominant ions are univalent, and the neutral species ($HCl°$, $NaCl°$, $NaOH°$) are in very low concentration (see Box, p. 39). The ionic strength for *the same concentration* of $CaCl_2$ on the other hand, is much greater because (1) the calcium ion is doubly charged, and (2) (therefore) there is twice as much chloride. This kind of I value is called *true* ionic strength because it is based on the species actually present in the solution. There is also the *stoichiometric* ionic strength, which assumes that all solutes are completely dissociated. A 1 molal NaCl solution therefore has a stoichiometric ionic strength of 1.00 molal, and a 1 molal $CaCl_2$ solution has a stoichiometric ionic strength of 3.00 molal.

Example   Ionic Strength

Speciation (see Chapter 6) of (1) a 0.1 molal NaCl solution, and (2) a 0.1 molal $CaCl_2$ solution by program SOLVEQ gives the following results:

0.1 m NaCl		0.1 m CaCl₂	
Species	Molality	Species	Molality
H^+	1.281×10^{-7}	H^+	1.727×10^{-7}
Cl^-	9.930×10^{-2}	Cl^-	1.869×10^{-1}
Na^+	9.930×10^{-2}	Ca^{2+}	8.705×10^{-2}
HCl°	1.118×10^{-9}	HCl°	2.491×10^{-9}
$NaCl^\circ$	7.023×10^{-4}	$CaCl^+$	1.281×10^{-2}
$NaOH^\circ$	1.494×10^{-9}	$CaCl_2^\circ$	1.408×10^{-4}
OH^-	1.277×10^{-7}	$CaOH^+$	6.887×10^{-8}
		OH^-	1.063×10^{-7}

The ionic strength is then

$$I = \frac{1}{2} \sum [(1)(0.0993) + (1)(0.0993) + \cdots]$$

all other species negligible, or zero charge

$$= 0.0993 \text{ molal, for NaCl}$$

and

$$I = \frac{1}{2} \sum \left[(1)(0.187) + (2^2)(0.0871) + (1)(0.0128) + \cdots \right]$$

$$= 0.274 \text{ molal, for CaCl}_2$$

The Davies Equation   For ionic strengths up to a few tenths molal, the Davies equation is reasonably accurate.

$$\log \gamma_{H_i} = \frac{-A z_i^2 \sqrt{I}}{1 + \sqrt{I}} + 0.2 A z_i^2 I \tag{3.19}$$

where A is a constant which varies slightly with temperature, and z_i is the charge on the ion (the valence). The 0.2 in the final term, being entirely empirical, is often changed to 0.3.

The B-dot Equation   Another approach is the "B-dot" equation. This has a long history, but geochemists generally use the version of Helgeson (1969),

$$\log \gamma_{H_i} = \frac{-A z_i^2 \sqrt{I}}{1 + B \mathring{a}_i \sqrt{I}} + \dot{B} I \tag{3.20}$$

Example   The Davies Equation

At 25°C, $A = 0.5091 \text{ kg}^{\frac{1}{2}} \text{ mol}^{\frac{1}{2}}$, so for the solutions from the previous example,

$$\log \gamma_{H,Na^+} = \frac{-0.5091 \times (+1)\sqrt{0.0993}}{1 + \sqrt{0.0993}} + [0.2 \times 0.5091 \times (+1) \times 0.0993]$$

$$= -0.112$$

$$\gamma_{H,Na^+} = 0.773$$

and

$$\log \gamma_{H,Ca^{2+}} = \frac{-0.5091 \times (2^2)\sqrt{0.274}}{1 + \sqrt{0.274}} + \left[0.2 \times 0.5091 \times (2^2) \times 0.274\right]$$

$$= -0.588$$

$$\gamma_{H,Ca^{2+}} = 0.258$$

Note that with the Davies equation, all singly charged ions will have the same activity coefficient, and all doubly charged ions will have another coefficient. This is not perhaps realistic, but the error introduced is often negligible compared with errors from other sources.

---

where B is another constant ($0.3283 \text{ kg}^{\frac{1}{2}} \text{ mol}^{-1} \text{ cm}^{-1}$ at 25°C), as long as $\mathring{a}_i$ is measured in angstroms (1 angstrom = 10^{-8} cm); $\mathring{a}_i$ is theoretically a distance of closest approach between ions of opposite charge, but is in practice an adjustable parameter. Values of $\mathring{a}_i$ for various ions (i) can be found in tables in books on physical chemistry, and are incorporated into many geochemical modeling programs. $\dot{B}$ is an empirical parameter designed to reproduce the activity coefficients of NaCl solutions, and therefore, (3.20) works best in solutions in which NaCl is the dominant solute. It will predict activity coefficients of Na^+ and Cl^- reasonably well to concentrations of perhaps 3 molal, and of other ions to 0.5 to 1 molal.

Uncharged Aqueous Species

As might be expected, measured activity coefficients for neutral (uncharged) aqueous species are generally fairly close to 1.0, although they are a function of solution concentration. However, only those neutral species which are directly measurable by chemical analysis (such as $SiO_2(aq)$, $H_2S(aq)$, $CO_2(aq)$) have had their activity coefficients measured. They are most commonly fitted to the ionic strength using the empirical *Setchénow equation*,

$$\log \gamma_i = k_s \cdot I \tag{3.21}$$

where in this case i is an uncharged solute (e.g., $SiO_2(aq)$), and k_s is a fit coefficient.

More difficult is the problem of the activity coefficients of uncharged metal complexes, present in low concentrations, and which may form only a small proportion of the metal in solution. These are usually assumed to have coefficients similar to the measured ones just referred to, without much justification.

Caveat Emptor   Modelers should realize that activity coefficients are affected by all solution components, and that simple, all-inclusive equations such as those above, which rely on the ionic strength (I) to work equally well for all compositions, cannot be expected to be very accurate. Moreover, the degree to which they are inaccurate in specific cases is usually not known. Undoubtedly, the situation is helped considerably by the fact that in equilibrium calculations the errors in calculated activity coefficients of products and reactants cancel one another to a large extent. Nevertheless, uncertainties in activity coefficients in geochemical models are always a major concern.

The Virial Equation Approach – The Pitzer Equations

Most other approaches to the calculation of activity coefficients for solution components, including solid and gaseous as well as liquid solutions, have used some form of a virial equation as a starting point. A virial equation is simply an equation for the ideal state (e.g., the ideal gas equation) followed by an ascending polynomial in one of the state variables. It seems to work well as a basis for activity coefficients because the form of the equation has a basis in statistical mechanics.

In the 1970s, Kenneth Pitzer and his associates developed a theoretical model for electrolyte solutions which combined the Debye–Hückel equation with additional terms in the form of a virial equation, which has proven to be extraordinarily successful at fitting the behavior of mixed-salt solutions to very high concentrations. This model has no provision for adjusting standard state parameters, or for considering individual reactions between species. It is also limited at the present time to relatively low temperatures and pressures. The equations involved are too lengthy and complex to present here, but they are used in some geochemical modeling programs. They should be considered when the environmental problem involves very concentrated solutions. See Anderson and Crerar (1993, §17.8) for an overview.

Solid and Gaseous Solutions

In both solid and gaseous solutions, virial equation-based Raoultian coefficients have often been proposed. For example, the Margules equations, often used in binary and sometimes in ternary solid solutions and which have a virial equation basis, were proposed originally for gaseous solutions. However, there is no satisfactory general model for Raoultian coefficients in multi-component solid solutions, and the tendency in modeling has been to treat these solutions as ideal (i.e., to use the mole fraction of a solid solution component as its activity; see Equation (3.13)).

Fugacity coefficients are important in considering the boiling of hydrothermal fluids, and have been approached from the virial equation as well as from numerous modifications of the van der Waals equation, the best known of these being the (modified) Redlich–Kwong equation. However, they are of minor importance in most environmental situations, and are routinely assumed to be 1.0, so that the activity of gaseous solution components is equal to the partial pressure (Equation (3.14)).

Satisfactory generalized equations for the calculation of activity coefficients in solid, liquid, and gaseous solutions under geological conditions will probably remain an important research goal for many years to come.

3.4.3   Chemical Potential

Equation (3.12) will often be seen in another form,

$$\mu_i - \mu_i^\circ = RT \ln a_i \tag{3.22}$$

where μ_i is substituted for G_i and is the *chemical potential* of i. μ_i is also the Gibbs energy per mole of i, but its definition (not explicitly given here) takes care of the fact that the magnitude of G_i varies with the concentration of i if i is a solute. It is therefore more correct to use μ when discussing the Gibbs energy of solutes, and it can be used for pure compounds as well as solutes. In other words, it is a more general way of expressing the Gibbs energy per mole of i, whatever i is. Combining this statement with our previous statement that a chemical reaction is balanced (at equilibrium) when the G (or μ) of reactants and the G (or μ) of products are equal, we find that for a generalized chemical reaction

$$a\mathrm{A} + b\mathrm{B} = c\mathrm{C} + d\mathrm{D} \tag{3.23}$$

to be at equilibrium, it is necessary that

$$\Delta_r G = \Delta_r \mu$$
$$= \underbrace{c\mu_\mathrm{C} + d\mu_\mathrm{D}}_{\text{products}} - \underbrace{a\mu_\mathrm{A} - b\mu_\mathrm{B}}_{\text{reactants}}$$
$$= 0 \tag{3.24}$$

3.5   The Equilibrium Constant

Because for each reactant and product in (3.23) (i.e., A, B, C, and D) there is a relation, from (3.22), such that

$$\mu_i = \mu_i^\circ + RT \ln a_i$$

it follows from (3.22) that

$$\Delta_r \mu - \Delta_r \mu^\circ = \Delta_r G - \Delta_r G^\circ$$
$$= RT \ln \frac{a_\mathrm{C}^c a_\mathrm{D}^d}{a_\mathrm{A}^a a_\mathrm{B}^b}$$
$$= RT \ln Q \tag{3.25}$$

Evidently $RT \ln Q$ is a term which measures the difference between $\Delta_r G^\circ$, the *tabulated* or standard state Gibbs energy of reaction, and $\Delta_r G$, the *real* Gibbs energy of reaction. When the activities are such that the real difference is zero ($\Delta_r G = 0$), the reaction is at equilibrium, and Q is called K. In this case,

$$\Delta_r G^\circ = -RT \ln K \tag{3.26}$$

where K is the *equilibrium constant*. Equation (3.26) is a remarkably powerful relationship. It says that for any reaction (for which we have data) we can use tabulated data for substances in their arbitrary reference or standard states to calculate the equilibrium relationship between product and reactant activities (concentrations) in our real systems.

For example, for our calcite dissolution reaction (3.2), we have the following data (from the EQ3 database):

Substance	$\Delta_f G^\circ$ (cal mol^{-1})
Ca^{2+}	$-132\ 120$
CO$_3^{2-}$	$-126\ 191$
CaCO$_3(s)$	$-269\ 880$

So

$$\Delta_r G^\circ = \Delta_f G^\circ_{CO_3^{2-}} + \Delta_f G^\circ_{Ca^{2+}} - \Delta_f G^\circ_{CaCO_3(s)}$$
$$= -126\ 191 - 132\ 120 - (-269\ 880)$$
$$= 11\ 569 \text{ cal mol}^{-1} \tag{3.27}$$

The fact that $\Delta_r G^\circ$ is positive means that the reaction *as written* will proceed to the left (calcite precipitates), if all products and reactants are in their standard states. Unfortunately, although pure solid calcite is in its standard state (and therefore has an activity of 1.0), the calcium and carbonate ions are *never* in their standard states, because this has been chosen to be a hypothetical ideal solution with a concentration of 1 molal (it is not surprising that calcite would be calculated to precipitate from a solution with the ions at 1 molal!). Therefore, this result is not very useful, except in the next step.

Using a value of $R = 1.98722$ cal K^{-1} mol^{-1}, $T = 298.15$ K, and the conversion factor $\ln x = 2.30259 \log x$, we use $\Delta_r G^\circ$ to calculate $\log K$, from (3.26), as

$$\log K = \frac{-11\ 569}{2.30259 \times 1.98722 \times 298.15}$$
$$= -8.480 \tag{3.28}$$

Thus the equilibrium constant can be calculated for any reaction for which we have standard state data for each product and reactant. Note that this does not necessarily mean that the reaction is important, or that it has reached equilibrium in our natural system, or perhaps any natural system: it tells us the ratio of product and reactant activities ($\approx$ concentrations) if the reaction *does* reach equilibrium.

3.5.1 Direct and Indirect Determination of K values

Determination of equilibrium constants from tables of Gibbs free energy values as described above is, in a sense, an *indirect* method, because the Gibbs energy values are themselves determined from other kinds of measurements, often calorimetric (measurements of quantities of heat involved in carefully controlled experiments). For the calcite reaction we have just considered, three separate free energy values are involved, and an error in any of them will result in an error in the equilibrium constant.

The equilibrium constants of many reactions can, however, be determined *directly* by measuring a solubility or gas pressure. In our calcite example, the equilibrium constant has been determined more or less directly by careful measurements of calcite solubility; e.g., Plummer and Busenberg (1982). The databases of geochemical models can be constructed to contain *either* $\Delta_f G°$ (e.g., SUPCRT92) or log K data (e.g., MINTEQA2, PHREEQC), and programs can solve for the equilibrium state of a system using free energies or equilibrium constants. Direct measurements of log K are often preferred (Nordstrom *et al.*, 1990) because of the likelihood that the quantity actually used in calculations (K or log K) will be less prone to errors.

Nevertheless, accurate values of $\Delta_f G°$ are inherently more useful than accurate values of log K, in the sense that log K refers to a single reaction, whereas a few values of $\Delta_f G°$ can be used in innumerable reactions, some of which may be difficult or impossible to measure directly. For the environmental modeler, this point may be a bit academic, because the modeler is responsible for ensuring the accuracy of the data used, in whatever form it occurs. In comparing data from various sources, a knowledge of the basic relationships described here is necessary to convert data from one form to the other.

3.5.2 Solubility Product and Saturation Index

Equilibrium constants for various kinds of reactions have been given various names. Reactions such as (3.2), with a solid phase on one side and its constituent ions on the other, is called a *solubility product reaction*, and the equilibrium constant for the reaction is called the *solubility product*, K_{sp}. In this case, we have found that $K_{sp}(3.2) = 10^{-9.971}$.

This means that, assuming that we are able to determine $a_{Ca^{2+}}$ and $a_{CO_3^{2-}}$ in a solution, we can then say whether calcite is undersaturated, supersaturated, or at equilibrium with the solution. This is useful because we quite often have samples of a groundwater but no information on the minerals in the host rock or soil it came from, and even if we know that calcite was or was not in the host formation, we could still not say whether it was dissolving or precipitating, or neither. The solubility product allows us to give a (theoretical) answer to this question. The determination of $a_{Ca^{2+}}$, $a_{CO_3^{2-}}$, and other species activities in a solution is one of the jobs of geochemical modeling programs, and will be discussed later on.

Note that Q and K are identical in form. The difference is that the activity terms in Q are not equilibrium activities, while those in K are. Similarly, if calcite is present, $a_{CaCO_3(s)} = 1$, and $K = K_{sp}$. If $Q = K_{sp}$, $\Delta_r \mu = 0$, and calcite is at equilibrium with its aqueous ions. When IAP > K_{sp}, $\Delta_r \mu > 0$, and calcite will precipitate. When

IAP, K_{sp}	Ω	SI ($= \log \frac{IAP}{K_{sp}}$)	Result
IAP $< K_{sp}$	< 1	negative	mineral dissolves
IAP $> K_{sp}$	> 1	positive	mineral precipitates
IAP $= K_{sp}$	1	0	equilibrium

Table 3.2. Relations between IAP, K_{sp}, and SI.

IAP $< K_{sp}$, $\Delta_r\mu < 0$, and calcite will dissolve. The quantity $(a_{Ca^{2+}} \cdot a_{CO_3^{2-}})$ in a real solution is called, appropriately enough, the *Ion Activity Product* (IAP) for calcite, and similarly for any other solubility product reaction. The IAP$/K_{sp}$ ratio is called Ω, and the logarithm of the ratio is called the *Saturation Index* (SI), so that when SI > 0 the mineral precipitates, and when SI < 0 the mineral dissolves (Table 3.2).

Example   IAP and SI

Speciation of sample 912-18 (at 25°C) in Merino (1975) using program SOLMIN88 gives the following results for calcium and carbonate (see Table 3.4):

Species	Molality	γ_H	Activity
Ca^{2+}	0.1681E-01	0.2623	0.4409E-02
CO_3^{2-}	0.1737E-04	0.2408	0.4183E-05

The IAP is therefore $(4.409 \times 10^{-3})(4.183 \times 10^{-6}) = 10^{-7.734}$. From Equation (3.28), $K_{sp} = 10^{-8.480}$. Thus

$$\begin{aligned} SI &= \log(IAP) - \log(K_{sp}) \\ &= -7.734 - (-8.480) \\ &= 0.746 \end{aligned}$$

which is positive, so calcite is supersaturated in this solution.

3.5.3   Dependence of K on Temperature

Combining equations (3.26) and (3.11), we obtain

$$\ln K = \frac{-\Delta_r H^\circ}{RT} + \frac{\Delta_r S^\circ}{R} \tag{3.29}$$

showing that if $\Delta_r H^\circ$ and $\Delta_r S^\circ$ are assumed to be independent of T, $\ln K$ will be a linear function of $1/T$. This is often a useful approximation over small temperature intervals. It is often used in geochemical modeling programs, where differences from ambient temperatures are usually small. If $\ln K$ is known at one temperature, say 298 K, it is easily shown that the same assumption allows calculation of $\ln K$ at another temperature T knowing only $\Delta_r H°$,

$$\ln K_T = \ln K_{298} - \frac{\Delta_r H°_{298}}{R} \left(\frac{1}{T} - \frac{1}{298.15} \right) \tag{3.30}$$

or

$$\log K_T = \log K_{298} - \frac{\Delta_r H°_{298}}{2.30259\, R} \left(\frac{1}{T} - \frac{1}{298.15} \right) \tag{3.31}$$

so that quite often almost the only data to be found in the database of a modeling program are values of $\log K$ and $\Delta_r H°$ for numerous reactions.

3.6 Components and Species

Some understanding of the concepts of species and components is essential to setting up geochemical models and interpreting their results.

3.6.1 Components and the Basis

If one looks up the term *component* in practically any text on physical chemistry or thermodynamics, one finds it is defined as the minimum number of chemical formula units needed to describe the composition of all parts of the system. We say *formulas* rather than *substances* because the chemical formulas need not correspond to any actual compounds. For example, a solution of salt in water has two components, NaCl and H_2O, even if there is a vapor phase and/or a solid phase (ice or halite), because some combination of those two formulas can describe the composition of every phase. Similarly, a mixture of nitrogen and hydrogen needs only two components, such as N_2 and H_2, despite the fact that much of the gas may exist as *species* NH_3. Note that although there is always a wide choice of components for a given system (we could equally well choose N and H as our components, or N_{10} and H_{10}), the *number* of components for a given system is fixed. The components are simply "building blocks", or mathematical entities, with which we are able to describe the bulk composition of any phase in the system. The list of components chosen to represent a system is, in mathematical terms, a basis vector, or simply "the basis".

3.6.2 Species

We immediately see a possible confusion between *components* and *species*. Both N_2 and H_2 can exist as physical *species* in the gas, as well as being the chosen abstract *components*. This confusion would not exist if we had chosen N and H as our components, because nitrogen does not exist in the system as N, only as diatomic N_2, and as NH_3. Perhaps unfortunately, components are commonly chosen to be entities which also exist as species, but there is a big difference between, say, component N_2 and species N_2 in the mixture of N_2 and H_2 gases.

Example   Components and Species

If a gas contains 1 mole of component N_2 and 3 moles of component H_2 ($n_{N_2} = 1$; $n_{H_2} = 3$) at 25°C, at equilibrium there will be 0.03177 moles of species N_2, 0.0953 moles of species H_2, and 1.9365 moles of species NH_3. Thus

$$n_{\text{component } N_2} = n_{\text{species } N_2} + \tfrac{1}{2} n_{\text{species } NH_3}$$
$$= 0.03177 + \tfrac{1}{2} \times 1.9365$$
$$= 1.00$$

and

$$n_{\text{component } H_2} = n_{\text{species } H_2} + \tfrac{3}{2} n_{\text{species } NH_3}$$
$$= 0.09539 + \tfrac{3}{2} \times 1.9365$$
$$= 3.00$$

Clearly, *component* N_2 represents the total amount of nitrogen in the system, while *species* N_2 represents nitrogen present in that exact stoichiometry in the system. The distinction is important, but normally the meaning of "N_2" is evident from the context.

3.6.3   An Alternative Basis

Aqueous geochemists, however, are interested not only in the "composition of all parts of the system", but also in the concentrations of all aqueous species in the system, including charged ions. But we cannot describe the concentration of Na^+ or H^+ ions in a salt solution using the formulas $NaCl$ and H_2O: no combination of the formulas $NaCl$ and H_2O will result in the formula Na^+. Furthermore, modelers want not only to "describe" the composition of systems, but to control, or *constrain*, their evolution during some process, such as maintaining equilibrium with some solid or gas phase.

Basis Species

To do these things, we need to use "building blocks" different from those we would choose to merely describe bulk compositions as illustrated above. We must use not only a different basis, but a different kind of basis. We could use the elements themselves (Al, B, N, ... , etc.), plus the electronic charge, because this would certainly allow us to describe the composition of any species or phase. However, it has proved to be convenient to use as "building blocks", or descriptive composition terms, entities which do exist – ordinary charged ions such as HCO_3^- and Na^+. These are called *basis species*, *component species*, or *master species*, and they make up a new kind of basis, which is the minimum number of chemical formulas needed to describe the composition of all phases *and all species, charged and uncharged* in the system. If mineral or gas phases are present, their compositions must also be included in the basis, as described below.

Example   Basis species

Suppose a system contains 1 mole of NaCl and 1 kg of water. The components as defined in §3.6.1 are NaCl and H_2O, but the *basis species* needed to describe all the ions present are (in most programs; other choices are always possible): Na^+, Cl^-, H^+, and H_2O. A speciation calculation gives the following results:

	Species	Molality
1	Cl^-	0.990056
2	Na^+	0.990056
3	$NaCl°$	0.009944
4	H^+	1.556E-07
5	OH^-	1.233E-07
6	$NaOH°$	3.231E-08
7	$HCl°$	6.085E-14

Note that the composition of all seven actual species can be described by some combination of the four basis species (e.g., $NaOH = Na^+ + H_2O - H^+$), and that each basis species (other than H^+ and H_2O) represents the total amount of some element. Thus

$$m_{\text{basis species } Cl^-} = m_{Cl^-} + m_{NaCl°} + m_{HCl°}$$
$$= 0.990056 + 0.009944 + 6.085 \times 10^{-14}$$
$$= 1.00$$
$$m_{\text{basis species } Na^+} = m_{Na^+} + m_{NaCl°} + m_{NaOH°}$$
$$= 0.990056 + 0.009944 + 3.231 \times 10^{-8}$$
$$= 1.00$$

Most modeling programs have a selection of 30 to 80 or more basis species, plus a collection of minerals and gases, from which the modeler chooses those required to describe the composition of all aqueous species, gases, and minerals in a particular system. If an element, say rubidium (Rb), does not occur as a basis species (as say, Rb^+) in the database, the program is of course then unable to calculate the amounts of various rubidium species or minerals, even if we have an analysis for the rubidium content of our system.

Auxiliary or Secondary Species

Having chosen a few basis species to be used in "building" or describing the composition of all other species and phases in the system, all other remaining species are called auxiliary or secondary species. In modeling programs there must, of course, exist some relationship between the two sets of species. Normally this consists of the stoichiometry of a secondary species in terms of basis species, plus the equilibrium constant of a reaction linking the two. Similarly, all minerals and gases included in the program must also be linked to the basis species by appropriate reactions and their equilibrium

Li	1.58	Ca	904	Br	132
Na	9470	Sr	58	I	40.0
K	151	Ba	0.70	alkalinity	1410
Rb	0.376	Fe	1.78	SO_4^{2-}	335
NH_3	44.0	F	0.50	H_2S	2.05
Mg	59	Cl	16 100	SiO_2	70
B	55	pH	6.8		

Table 3.3. Analysis of sample 912-18 from the Kettleman North Dome oil field (Merino, 1975). Data are in mg L^{-1}.

constants. Examples of these "links" are given in §4.3.1.

The distinction, then, between a species which actually exists in the real system, say, the sodium ion Na^+, and the *basis species*, Na^+, is very important. Just as in the nitrogen example above, *component* Na^+ represents the total amount of sodium in the system, and *species* Na^+ represents the sodium actually present as the univalent sodium ion in the solution. Similarly, in the output from the program, basis species and "real" species are commonly mixed together in some way, which is quite clear only if we are perfectly aware of the difference.

An Example   For example, the analytical data for sample 912-18 from Merino (1975), referred to earlier, are shown in Table 3.3, and part of the output file from SOLMIN88 is shown in Table 3.4. (Note, in passing, that the data for Ca^{2+} and CO_3^{2-} from this Table were used in the example on page 45.)

There are 16 species which have numbers in the two "Analyzed" columns, one for each of the species analyzed (Table 3.3) except for Rb, Br, and I, which are ignored by SOLMIN88 because it has no basis species (no data) for these elements. The data for "alkalinity" turn up as data for HCO_3^- (this is under user control during input), and the pH is entered directly in the "$-\log_{10}$ activity" column (omitted from Table 3.4). Note too that, because the basis species for most elements is the same as the component reported in the analysis, the "mg/L" column contains the analytical numbers from Table 3.3. The exception is boron, which is reported in Table 3.3 as 55 mg L^{-1} B, but as 314.6 mg L^{-1} $B(OH)_3$ in the program output. In other words, the basis species chosen for boron is not B but $B(OH)_3$, and the program has made the conversion by multiplying 55 by the ratio of the molecular weights of $B(OH)_3$ and B; thus $55 \times 61.833/10.811 = 314.6$.

It seems reasonable, then, that the elements actually analyzed appear in the "Analyzed" columns in the program output. What may be confusing is that many of them appear as both "Analyzed" and "Calculated", and that the numbers in these two categories are completely different. For example, Ca^{2+} is analyzed at 904 mg L^{-1}, but is calculated to be 654.3 mg L^{-1}. It must be understood that in the "Analyzed" columns, Ca^{2+} represents the basis species chosen for calcium; it is the calcium *component*, which equals the total calcium content of the solution. In the "Calculated" columns however, Ca^{2+} represents one of the calcium *species* actually present in the solution.

| | SPECIES | ---ANALYZED---------- | | -------------CALCULATED---------- | | | | ACTIVITY |
		MG/L	MOLALITY	PPM	MG/L	MOLALITY	ACTIVITY	COEFF.
1	Ca ++	904.0000	0.2323E-01	0.6543E+03	0.6543E+03	0.1681E-01	0.4409E-02	0.2623
2	Mg ++	59.0000	0.2499E-02	0.5392E+02	0.5392E+02	0.2284E-02	0.6385E-03	0.2796
3	Na +	9470.0000	0.4242E+00	0.9170E+04	0.9170E+04	0.4108E+00	0.2801E+00	0.6819
4	K +	151.0000	0.3977E-02	0.1474E+03	0.1474E+03	0.3881E-02	0.2496E-02	0.6432
5	Cl -	16100.0000	0.4676E+00	0.1557E+05	0.1557E+05	0.4522E+00	0.2909E+00	0.6432
6	SO4 --	335.0000	0.3591E-02	0.1932E+03	0.1932E+03	0.2071E-02	0.3897E-03	0.1881
7	HCO3 -	1410.0000	0.2380E-01	0.1202E+04	0.1202E+04	0.2029E-01	0.1417E-01	0.6987
8	H +			0.1937E-03	0.1937E-03	0.1979E-06	0.1585E-06	0.8010
9	OH -			0.1581E-02	0.1581E-02	0.9575E-07	0.6353E-07	0.6635
11	H4SiO4			0.1118E+03	0.1118E+03	0.1198E-02	0.1604E-02	1.3392
12	SiO2	70.0000	0.1200E-02					
15	Ba ++	0.7000	0.5249E-05	0.6109E+00	0.6109E+00	0.4581E-05	0.1036E-05	0.2261
18	Fe ++	1.7800	0.3282E-04	0.1682E+01	0.1682E+01	0.3101E-04	0.8133E-05	0.2623
22	Li +	1.5800	0.2345E-03	0.1577E+01	0.1577E+01	0.2341E-03	0.1734E-03	0.7410
26	Sr ++	58.0000	0.6817E-03	0.5342E+02	0.5342E+02	0.6278E-03	0.1419E-03	0.2261
30	F -	0.5000	0.2710E-04	0.4708E+00	0.4708E+00	0.2552E-04	0.1693E-04	0.6635
31	B(OH)3	314.6000	0.5239E-02	0.3125E+03	0.3125E+03	0.5205E-02	0.5831E-02	1.1204
32	NH3	44.0000	0.2660E-02					
33	H2S	2.0500	0.6194E-04	0.1023E+01	0.1023E+01	0.3091E-04	0.3185E-04	1.0305
51	BaCO3			0.2344E-03	0.2344E-03	0.1223E-08	0.1370E-08	1.1204
52	BaHCO3			0.1142E+00	0.1142E+00	0.5930E-06	0.4044E-06	0.6819
53	BaOH +			0.4680E-07	0.4680E-07	0.3123E-12	0.2230E-12	0.7140
54	BaSO4			0.1667E-01	0.1667E-01	0.7356E-07	0.8241E-07	1.1204
55	CaCO3			0.2655E+01	0.2655E+01	0.2732E-04	0.3061E-04	1.1204
56	CaHCO3 +			0.1067E+03	0.1067E+03	0.1086E-02	0.8051E-03	0.7410
57	CaOH +			0.2893E-03	0.2893E-03	0.5218E-08	0.3866E-08	0.7410
61	CaSO4			0.4045E+02	0.4045E+02	0.3060E-03	0.3428E-03	1.1204
71	FeCl +			0.9077E-01	0.9077E-01	0.1024E-05	0.6982E-06	0.6819
72	FeCl2			0.4662E-09	0.4662E-09	0.3787E-14	0.4243E-14	1.1204
75	FeOH +			0.1619E-02	0.1619E-02	0.2288E-07	0.1634E-07	0.7140
76	Fe(OH)2			0.6276E-07	0.6276E-07	0.7192E-12	0.8058E-12	1.1204
77	FeOOH -			0.2390E-08	0.2390E-08	0.2770E-13	0.1978E-13	0.7140
78	FeSO4			0.1123E+00	0.1123E+00	0.7614E-06	0.8531E-06	1.1204
90	H2SiO4 --			0.7150E-05	0.7150E-05	0.7825E-10	0.1884E-10	0.2408
91	H3SiO4 -			0.2027E+00	0.2027E+00	0.2195E-05	0.1497E-05	0.6819
97	H2CO3			0.2704E+03	0.2704E+03	0.4489E-02	0.5029E-02	1.1204
98	CO3 --			0.1012E+01	0.1012E+01	0.1737E-04	0.4183E-05	0.2408
281	CaCl +			0.3628E+03	0.3628E+03	0.4946E-02	0.3373E-02	0.6819
282	CaCl2			0.5308E+01	0.5308E+01	0.4925E-04	0.5517E-04	1.1204

Table 3.4. Some of the speciation results for sample 912-18 (Merino, 1975), as produced by program SOLMIN88. The original printout also includes columns "PPM" in the "ANALYZED" section and "log10 activity" in the "CALCULATED" sections. These have been removed to allow the data to fit the page.

Other calcium species are listed further down, and have numbers only in the "Calculated" columns. The sum of the molalities of all Ca-containing species is equal to the input value of 0.02323 molal.

Another example of this important difference is provided by the results for silica. The analyzed silica content (Table 3.3) is 70 $mg\,L^{-1}$, and this appears as the basis species SiO_2. This has been converted by the program into a molality of 0.0012. In many applications, $SiO_2(aq)$ and H_4SiO_4 are used synonymously as the uncharged monomeric aqueous silica *species*. But here, SiO_2 is the total silica and the "Calculated" amount of H_4SiO_4 is 0.001198, a bit less. This is because the basis species SiO_2 has been split up by the program into the three secondary species H_4SiO_4, $H_3SiO_4^-$, and $H_2SiO_4^{2-}$, seen further down in the results. The sum of the molalities of these three species is equal to the input value, 0.0012 molal.

3.7   The Phase Rule

We now seem to have two types of *components*. For example, for the system NaCl–H_2O, we have either the two "traditional" components NaCl and H_2O, which allow us to describe the bulk composition of all phases in this system, or we have the four *basis species* Na^+, Cl^-, H^+, and H_2O, which allow us to describe not only the compositions of the phases but also the concentration of all dissolved species in the system. "Traditional" components and basis species are simply different choices of components, which have different purposes and different descriptive powers. We need more basis species because they are called upon to provide more information.

Readers familiar with Morel and Hering (1993) will recognize that what we have termed "traditional" components are Morel and Hering's "recipes", and what we refer to as basis species they call simply components.

Degrees of Freedom

The Phase Rule links the number of components and the number of phases present at equilibrium to something called *degrees of freedom*. The number of degrees of freedom possessed by a system is the number of properties of the system which must be specified to specify or fix completely the equilibrium state of the system. This number is of some importance to modelers, as it is the number of pieces of information about a system which must be supplied to a modeling program before it can begin.

The concept of degrees of freedom is perhaps most easily seen in mathematical terms. It boils down to the fact that if we have a mathematical "system" consisting of n unknown variables, we need n relationships or equations among the variables in order to determine all the variables. If there are less than n relationships or equations, then there are some degrees of freedom; we have to supply some information before we can solve for all the variables in the system.

For example, if the system consists of three variables x, y, and z, and we have no information whatsoever about them, we have three degrees of freedom. If we have one equation relating the variables, such as $2x + 3y + z = 0$, then we have two degrees of freedom, meaning that if we supply the values of any two of the variables, the third will be determined by the equation. If we have two equations, we need to supply the value of only one variable, and the system has one degree of freedom. If we have three equations, then we can solve for all variables, there are zero degrees of freedom, and the system is "invariant". In other words, the number of degrees of freedom is equal to the number of variables or unknown quantities, minus the number of known relationships or equations among them.

The Phase Rule is simply an application of this fundamental principle to chemical systems. It can be derived from fundamental thermodynamic equations in the manner described above, but we will describe it here in more intuitive terms.

Traditional Components   For example, consider our two-component system NaCl–H_2O. If there is one homogeneous liquid phase, there are three degrees of freedom. We must know the temperature, pressure, and the concentration of NaCl in order to define the system completely. In theory, we could specify *any* three system properties, but these three are the most commonly used. Any two is too few, and four would be redundant.[1]

However, if there are two phases, say a salt solution and its coexisting vapor phase, then there are only two degrees of freedom, because only two properties need to be specified. If we specify the temperature and the salt concentration, then we have no choice of pressures – pressure is fixed by the vapor pressure of the solution. And if there are three phases, we have only one degree of freedom – we can choose a salt concentration, for example, but an ice phase or a halite phase plus a vapor phase can only coexist at one T and P for that concentration.

In other words, the more phases there are, the fewer properties we need to specify to describe the equilibrium state. The Phase Rule sums all this up as

$$f = c - p + 2 \tag{3.32}$$

where f is the number of degrees of freedom, c is the number of components, and p is the number of phases.

Of course, as we habitually consider systems at some fixed value of T and P, this "uses up" two degrees of freedom, and so if T and P are understood to be already chosen (as is commonly the case in environmental systems, where the temperature and pressure have their normal ambient values), the Phase Rule can be re-written

$$f = c - p \tag{3.33}$$

Basis Species   If, however, we use basis species as components, we will have more degrees of freedom to deal with. For example, using components NaCl and H_2O, we have no control over the Na/Cl ratio, but using basis species Na^+, Cl^-, H^+, and H_2O, we do – we can specify Na^+ and Cl^- independently – an extra degree of freedom. In this case, the Phase Rule is no different, but we change the notation. Phase Rule (3.33) becomes, for a system having a specified T and P,

$$f = b - p \tag{3.34}$$

where b is the number of basis species needed to define the system; in this case $b = 4$.

Fortunately, this is fairly intuitive. It just says that to define an aqueous solution ($p = 1$) at a given T and P, we have to specify the concentration of each solute element. That is, since H_2O is always one of the basis species, then there are $(b - 1)$ degrees of freedom, which is evidently the number of *solute* basis species (Na^+, Cl^-, and H^+). Each additional phase present fixes the value of one basis species, and hence reduces by one the number of basis species that must be specified.[2] These additional phases must be incorporated into the basis, as described in §5.8.1.

---

[1] Note too that by "defining the system" we mean defining the intensive parameters – the masses or volumes of the various phases or components are irrelevant in this traditional use of the Phase Rule. A gram of NaCl solution is no different from a kilogram of solution in this sense. We will consider what happens if extensive parameters are included in §3.7.1.

[2] For example, if halite was present at equilibrium, we would specify either the Na^+ or the Cl^- basis species, but not both. The other would be fixed through the solubility product. Thus an additional phase always reduces the number of independent basis species by one.

Charge Balance   These basis species are actually not completely independent of one another, as theoretically required, because the sum of the positive and negative charges must be equal – there must be a charge balance in real solutions. However, the charge balance requirement does not really reduce the degrees of freedom. It just means that one of the charged basis species, say Cl^-, should be adjusted to give the balance. The charge balance requirement is thus substituted for the requirement that basis species Cl^- be specified.

Gas Phases   In many environmental problems, gas phase compositions are not very important. Most programs reflect this in treating gases in a simple way. Each is considered individually, rather than as members of a single gaseous solution. Therefore, each gas which is controlled, or buffered, in model systems is effectively a separate phase.

3.7.1   The Extensive Phase Rule

Each of the Phase Rules above is used to "define the equilibrium state", which means that they each relate the number of *properties* (understood to be *intensive variables*) of the system to the number of degrees of freedom. This "defines" the equilibrium state, but it does not define *how much* of the equilibrium state we have. The "equilibrium state" of 1 kg of water saturated with halite is the same whether we have 1 g or 1 kg of halite. But modeling programs commonly want to do more than to define the equilibrium state. They want to dissolve or precipitate phases during processes controlled by the modeler, and to keep track of the masses involved, so as to know when phases should appear or disappear. To do this, the mass of each phase is required, not just its presence or absence. Therefore, an additional piece of information is required for each phase present, or p quantities. Almost invariably, the mass of H_2O is chosen as 1 kg, so that the concentration of basis species defines the mass of each.[3] If solid or gas phases are specified, the mass is usually also specified. If we count these extra p pieces of data, the *extensive* Phase Rule becomes

$$f = b \qquad\qquad (3.35)$$

This relationship is also fairly intuitive. Look at it this way. The number of phases is always at least one (a system with no phases is not very interesting). To define a system having only an aqueous solution phase ($p = 1$), we must specify each of the solutes in the water, or $b - 1$ quantities. If there is one mineral in equilibrium with the water ($p = 2$), it controls one basis species, and so reduces b by one, and similarly for all p mineral or gas phases. This is Phase Rule (3.34). But defining the equilibrium state is not usually enough. We want also to know the mass of each phase, so we need p extra data, giving Phase Rule (3.35), which says that for any system we need b pieces of information. These b pieces of information are

- the mass of water (almost always 1 kg);

---

[3]The unit of concentration used in modeling calculations is invariably molality, or the moles of solute species per kilogram of pure water. Therefore, if the mass of water is fixed at 1 kg, the molality of a species automatically equals the number of moles of the species, which is readily convertible to grams.

- the mass of each mineral or gas phase;

- the concentration of basis species beyond those controlled by the mineral and gas phases.

An example will help to clarify this (see the Box on page 55). However, if one finds this confusing, you are not alone, and it is not terribly important because most modelers do not actually start with the Phase Rule in constructing models, in spite of the fact that it does tell us how many data are required. If the modeler was starting from scratch, it would undoubtedly be useful, but when using well established programs, it is often more efficient to bypass the Phase Rule, and rely on error messages from the modeling program to get things right. It is useful, though, to know that the Phase Rule used or implied in geochemical modeling is somewhat different from the one derived by Gibbs.

Duhem's Theorem

The term "Extensive Phase Rule" is our own terminology, and may prove confusing to geochemists more used to seeing it referred to as Duhem's Theorem. As expressed by Prigogine and Defay (1965), p. 188, Duhem's Theorem says

> Whatever the number of phases, of components or of chemical reactions, the equilibrium state of a closed system, for which we know the initial masses $m_1^\circ \ldots m_c^\circ$, is completely determined by two independent variables.

By "completely determined", the masses of all phases is meant to be included. In the general case, where the system is not invariant or univariant, the "two independent variables" can be T and P, and this amounts to saying (from the Theorem) that in addition we need to know the mass of every traditional component, if we are not concerned with ionic speciation, or every basis species, if we are.

3.8 Redox

Many elements in natural systems occur in more than one state of oxidation, or valence state. Thus iron occurs as either Fe^{2+} or Fe^{3+}; arsenic occurs as either As^{3+} or As^{5+}; sulfur occurs in many valence states between S^{2-} and S^{6+}; and so on. If all these valence states have concentrations that are independent of one another, then each represents an additional component, and each must be specified or constrained somehow in setting up a geochemical model. However, if chemical equilibrium prevails in the system, these states are *not* all independent – each is dependent on the oxidation–reduction (redox) state of the system. Specifying this redox state is sufficient to specify the ratio of the activities of each pair of valence states, e.g., $a_{Fe^{3+}}/a_{Fe^{2+}}$, $a_{As^{5+}}/a_{As^{3+}}$, and so on. This ratio, together with the total amount of the element (total Fe, total As) in the system, is then sufficient to determine the activities of each species separately.

Therefore, in setting up equilibrium models of systems containing elements which occur in more than one valence state, one additional parameter is required – a measure of the redox state of the system. In some programs, the user is also allowed to specify a

---

Example **Phase Rules**

Consider a system consisting of water and dissolved Al, Si and K. If these solutes are sufficiently concentrated, aluminosilicate minerals will precipitate. Let's say that our T and P have been specified as 25°C and 1 bar.

- "Traditional" components for this system would be K_2O, Al_2O_3, SiO_2, and H_2O, so $c = 4$. If the solutes are very dilute and we have only one phase (water), Phase Rule (3.33) says $f = c - 1 = 3$, so we have to specify the concentrations of K, Al and Si to define the system. But if we have three solid phases in equilibrium with the water, such as kaolinite, muscovite, and quartz, then $f = c - p = 0$, and the system is invariant (we don't have to specify anything). However, even though all properties of the system are fixed, including the species in the liquid phase, we are unable to describe or calculate the ionic species using these components.

- Alternatively, we choose as the basis the five basis species (or component species) H_2O, K^+, Al^{3+}, H^+, and $SiO_2(aq)$, thus $b = 5$. With only water present, Phase Rule (3.34) says that we need to specify $f = 5-1 = 4$ concentrations, which could be the total amounts of K, Al, Si, plus the pH. With kaolinite, muscovite and quartz present in any amount at equilibrium, Phase Rule (3.34) says that $f = 5 - 4 = 1$, so we need to specify only one concentration to be invariant. In either case, i.e., whether we have specified one or four concentration(s), we *are* able to describe and calculate all ionic species present.

- Extensive Phase Rule (3.35) then says that even though we are invariant, that is, even though we have adequately described the equilibrium state, to make our description more useful we need $f = b = 5$ pieces of information: either (with only water) the four concentrations and the mass of water, or (with water and three minerals) one concentration plus the masses of four phases.

---

redox state, but also to "decouple" any redox pairs that may not have reached equilibrium. For example, we might specify the oxidation state of a system, but also specify the activities of Fe^{2+} and of Fe^{3+} independently, after "decoupling" this pair.

There are three principal ways of stating the redox state of a system.

3.8.1 Oxygen Fugacity, log f_{O_2}

The change in valence of an atom obviously involves the gain or loss of electrons. For example,

$$Fe^{2+} = Fe^{3+} + e \tag{3.36}$$

However, electrons do not simply float around in aqueous solutions in a freely available state. Electrons must be transferred to or from some other species which is willing to give them up or to accept them. Often, but not necessarily, this species is oxygen. Thus the ferrous/ferric reaction might be

$$Fe^{2+} + \tfrac{1}{4}O_2(aq) + H^+ = Fe^{3+} + \tfrac{1}{2}H_2O \tag{3.37}$$

In this reaction, the electron lost by Fe^{2+} is transferred to oxygen, which combines with H^+ to form water.

Of course, not all solutions contain dissolved oxygen, and many species other than oxygen may be the electron acceptor/donor. For example, if copper is also in the system, we might have

$$Fe^{2+} + Cu^{2+} = Fe^{3+} + Cu^+ \tag{3.38}$$

in which the electron lost by Fe^{2+} is accepted by Cu^{2+}, reducing it to Cu^+.

However, no matter how the electrons are transferred in reality, we can always *write* the electron transfer as in Equation (3.37), and use the activity of oxygen as an indicator or index of redox conditions, regardless of whether oxygen is actually in the system. This simply emphasizes the difference between real and model systems that we mentioned in §3.1.1.

As a numerical example, consider the reaction between the reduced and oxidized forms of sulfur:

$$H_2S(aq) + 2\,O_2(g) = SO_4^{2-} + 2\,H^+ \tag{3.39}$$

for which the equilibrium constant is

$$\frac{a_{SO_4^{2-}}\,a_{H^+}^2}{a_{H_2S(aq)}\,f_{O_2}^2} = K_{3.39} \tag{3.40}$$

If we look up the values of $\Delta_f G^\circ$ for each of these four species and calculate K, we will find it is about 10^{126} at 25°C. If we now choose a pH of 7.0 ($a_{H^+} = 10^{-7}$) and rearrange Equation (3.40), we find

$$f_{O_2} = 10^{-70} \left(\frac{a_{SO_4^{2-}}}{a_{H_2S(aq)}} \right)^{\frac{1}{2}} \tag{3.41}$$

Thus f_{O_2} is a quantity which is proportional to the sulfate–sulfide ratio. If we choose conditions such that $a_{SO_4^{2-}} = a_{H_2S(aq)}$, we find $f_{O_2} = 10^{-70}$ bars. Considered as a partial pressure of oxygen, this doesn't make much sense, as it would correspond to something like one molecule of oxygen in the entire universe. So the fugacity of oxygen, while it does approximate a partial pressure under some conditions, is better thought of as simply a thermodynamic parameter which is a useful index of the oxidation state of aqueous solutions (or almost any other system) under all conditions. In the above case, if the sulfate–sulfide ratio is 100 (solution more oxidized), the f_{O_2} is 10^{-69}; if the sulfate–sulfide ratio is 0.01 (solution more reduced), the f_{O_2} is 10^{-71}. Of course the same relationships hold for any redox pair, such as Fe^{3+}/Fe^{2+}; CO_2/CH_4; U^{6+}/U^{4+}; and so on. Each ratio may be different, but each pair can appear in a balanced reaction involving $O_2(g)$, and, if the solution is at equilibrium, each pair will result in the same f_{O_2}. Note that in this example, we were careful to use the labels (g) and (aq), because

$O_2(aq)$ and $O_2(g)$ are thermodynamically and physically quite different things, and similarly for $H_2S(aq)$ and $H_2S(g)$.

So we see that although f_{O_2} is a useful indicator of redox conditions, it becomes very small under reducing conditions. In these cases it is often more convenient to use hydrogen instead of oxygen as the indicator substance. The fugacities f_{O_2} and f_{H_2} are related in aqueous systems by the reaction

$$2\,H_2O(l) = O_2(g) + 2\,H_2(g) \tag{3.42}$$

for which the equilibrium constant is

$$K_{3.42} = \frac{f_{O_2} \cdot f_{H_2}^2}{a_{H_2O(l)}} \tag{3.43}$$

Therefore in aqueous systems, in which $a_{H_2O(l)} \approx 1$, the oxygen and hydrogen fugacities are inversely related, so that as f_{O_2} becomes very small, f_{H_2} (or a_{H_2}) becomes large, and in many cases is actually a measurable quantity, i.e., a partial pressure (or a molality).

3.8.2   Redox Potential, *Eh*

Redox reactions always involve a transfer of electrons. Reactions such as Equation (3.38) which show both the donor (Fe^{2+}) and the acceptor (Cu^{2+}) of electrons can always in theory form the basis of an electrochemical cell, and can be called cell reactions. They can always be broken down into their separate "half-cell reactions". For reaction (3.38), these are

$$Fe^{2+} = Fe^{3+} + e \tag{3.44}$$

and

$$Cu^{2+} + e = Cu^+ \tag{3.45}$$

and, with the aid of a few simple conventions, each of these half-cells can be assigned a voltage, either for standard conditions and unit activities ($\mathcal{E}^\circ$), or for actual conditions and real activities ($\mathcal{E}$).

A complete description of the conventions and calculations involved is too lengthy for inclusion here. The point is, however, that the half-cell potential, or voltage, of reactions such as (3.44) and (3.45) is directly proportional to the ratio of the product and reactant activities in the reaction. Thus, under standard conditions,

$$Fe^{2+} = Fe^{3+} + e \qquad \mathcal{E}^\circ = 0.769\,V \tag{3.46}$$

and, if the $a_{Fe^{2+}}/a_{Fe^{3+}}$ ratio happened to be 100,

$$Fe^{2+} = Fe^{3+} + e \qquad \mathcal{E} = 0.651\,V \tag{3.47}$$

This half-cell voltage (0.651 V) is called the redox potential of the fluid, *Eh*, thus both *Eh* and f_{O_2} are quantities which are proportional to the ratio of reduced to oxidized species in solution, and either may serve as an index of how oxidizing or reducing a solution is.

The conversion between Eh and $\log f_{O_2}$ at 25°C is

$$Eh = 1.23 + 0.0148 \log f_{O_2} - 0.0592 \, p\text{H} \qquad (3.48)$$

3.8.3 Electron Potential, pe

The concept of electron potential was introduced as an analogy between $p\text{H}$, which refers to hydrated protons, and pe, which would refer to hydrated electrons. Like $p\text{H}$, pe is defined in terms of activity:

$$p\text{H} = -\log a_{\text{H}^+}$$

$$pe = -\log a_{\text{e}} \qquad (3.49)$$

In fact the "p" notation is now widely used for various quantities. For example, equilibrium constants are sometimes given in terms of pK, where

$$pK = -\log \text{K}$$

In other words if $K = 10^{-6.37}$, then $pK = 6.37$.

The conversion between pe and Eh is given by

$$
\begin{aligned}
pe &= Eh(\mathcal{F}/2.30259 \, RT) \\
&= 5040 \, Eh/T \\
&= Eh/0.05916 \text{ at } 298.15 \text{ K} \qquad (3.50)
\end{aligned}
$$

where $\mathcal{F}$ is the Faraday constant, 96485.309 Coulombs mol^{-1}, or J V^{-1} mol^{-1}.

There are no hydrated electrons in real solutions, just as there may be no oxygen. However, this does not detract from the usefulness of pe or of $\log f_{O_2}$ as indices of redox conditions. They are calculated numbers, based on actual voltages or concentrations.

3.9 Alkalinity

In general terms, alkalinity is the "Acid Neutralizing Capacity" of a solution, that is, the quantity of acid required to "neutralize" the solution. The acidity is similarly the "Base Neutralizing Capacity", the quantity of base required to "neutralize" the acidity of a solution. Alkalinity and acidity are determined by titrating a sample of solution with an acid (such as HCl) or a base (such as NaOH) of known concentration. However, the variety of ways in which these simple concepts can be defined and interpreted has led to much confusion. Several modeling programs now do not allow input of acidity or alkalinity, as such, partly because of this confusion. However, others do, and in any case users will still have to deal with these concepts if they appear in their analyses.

3.9.1 The Carbonate Component

Whatever the historical reasons for these measurements, their principal use in geo-chemical modeling is as an indication of the carbonate content of the solution. The total carbonate content, or in our terms the carbonate component or basis species, is the quantity required in modeling. Therefore the discussion in this section is directed toward how to determine the total carbonate content of a solution, given an alkalinity, an acidity, or both. The carbonate content of solutions is generally one of the more important parameters in geochemical modeling. It is important to understand how to get whatever information we have about it into our program.

As mentioned in §3.6.3, the component or basis species chosen for most elements is usually the most representative or most common ionic form, thus Na^+ for sodium, SO_4^{2-} for sulfur, and so on. Sometimes, though, there are differences among programs. For the carbonate component, all possible choices are used in various programs. That is, the carbonate component, or total carbonate, may be represented by $CO_2(aq)$, H_2CO_3, HCO_3^-, or CO_3^{2-}. To avoid this confusion, we will refer to it here as $m_{CO_3,total}$. If this total carbonate is 10^{-3} m, so that

$$m_{CO_3,total} = m_{CO_3^{2-}} + m_{HCO_3^-} + m_{H_2CO_3^\circ} + m_{CO_2(aq)} \qquad (3.51)$$
$$= 10^{-3}$$

then Figures 3.1 and 3.2 sum up these relationships.[4] If total carbonate is not 10^{-3} m, the diagrams look exactly the same, but with other numbers on the y-axis.

3.9.2 Carbonate Speciation

In acidic solutions, below a pH of about 6.4, most dissolved carbonate exists as $H_2CO_3^\circ$ and $CO_2(aq)$, that is, hydrated and non-hydrated aqueous carbon dioxide, and there is an equilibrium relationship between them, the non-hydrated form being the dominant species. However, the distinction between these two forms is unimportant for most purposes, and although notation varies it has become fairly common to denote the total dissolved CO_2, whether hydrated or non-hydrated, as $H_2CO_3^*$. Thus,

$$m_{H_2CO_3^*} = m_{H_2CO_3^\circ} + m_{CO_2(aq)} \qquad (3.52)$$

Between pH values of 6.4 and 10.3, most carbonate is in the bicarbonate form, HCO_3^-, and above pH 10.3, most is in the carbonate form, CO_3^{2-}. These relationships follow directly from the dissociation reactions

$$H_2CO_3^* = HCO_3^- + H^+; \quad K = 10^{-6.37} \qquad (3.53)$$
$$HCO_3^- = CO_3^{2-} + H^+; \quad K = 10^{-10.33} \qquad (3.54)$$

---

[4]Note that in calculating these diagrams, it is necessary to use an acid and a base such as HCl and NaOH to vary the pH. Sodium ions form ion pairs with carbonate species to a small extent. These are not shown.

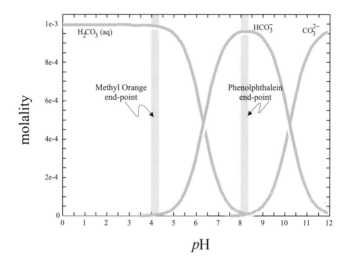

Figure 3.1. Molalities of $H_2CO_3^*$, bicarbonate, and carbonate ions as a function of pH, for a total concentration of 10^{-5} m. Sodium carbonate species not shown.

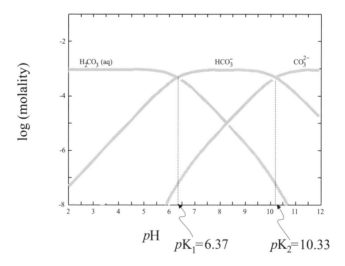

Figure 3.2. Same as Figure 3.1, but with a log molality scale. The pH values of the crossover points are approximately the pK values of carbonic acid ionization, Equations (3.53) and (3.54) (very slightly different due to activity coefficient effects).

3.9.3   Titration Alkalinity

Titrations are carried out to "end-points", defined by inflection points in titration curves. These theoretical end points are often difficult to determine in practice, so "fixed end-points" are now universally used, defined by the change in color of an indicator substance, or by a pH reading. For alkalinity titrations, a pH of about 4.3 is used, as indicated by an electrode or by an indicator such as methyl orange. In other words, any solution having a pH greater than 4.3 is titrated with a standard acid, and the quantity required to lower the pH to 4.3 is reported in milliequivalents per liter of solution (meq L^{-1}).[5] Alternatively, this quantity of milliequivalents may be converted into an equivalent quantity of calcite ($CaCO_3$), or of HCO_3^-.

Carbonate Solutions

Let's consider first natural solutions in which carbonate ions are dominant, such as most unpolluted fresh waters. As long as the pH is not less than 4.3, we can determine the amount of acid required to lower the pH to this value. The dominant reactions using up H^+ ions will be

$$HCO_3^- + H^+ = H_2CO_3^* \tag{3.55}$$

and

$$CO_3^{2-} + 2\,H^+ = H_2CO_3^* \tag{3.56}$$

In very basic or very dilute solutions, the OH^- content may also be significant.[6]

It follows from these relationships that we can define carbonate alkalinity as

$$\text{alkalinity} = m_{HCO_3^-} + 2\,m_{CO_3^{2-}} + m_{OH^-} - m_{H^+} \tag{3.57}$$

After determining the number of moles of acid used during the titration, this may be converted into an equivalent number of moles of $CaCO_3$ or HCO_3^-, and reported as "alkalinity, as $CaCO_3$", or "alkalinity, as HCO_3^-", or unfortunately, perhaps just as "alkalinity".

Non-carbonate Alkalinity

Solutions involved in environmental problems very commonly have a number of solutes other than carbonate, and many anions other than carbonate or bicarbonate can also neutralize acid during the titration. For example, solutions with significant amounts of

---

[5]For HCl, this is the same as millimoles per liter. For H_2SO_4, it is twice the number of millimoles per liter.

[6]Alternatively, following acidification the solution can be boiled, driving off the dissolved CO_2, which can then be collected, measured, and reported directly, which is one explanation of why "alkalinity" does not always occur in chemical analyses.

Species		
$CaCO_3$	HCO_3^-	CO_3^{2-}
100.0894	61.0174	60.0174

Table 3.5. Carbonate molecular weights (NBS Tables, Wagman *et al.*, 1982).

phosphate, boron, bisulfide, silicate, ammonia, or organic anions can use up H^+ through the reactions

$$HPO_4^{2-} + H^+ = H_2PO_4^- \tag{3.58}$$

$$B(OH)_4^- + H^+ = B(OH)_3 + H_2O \tag{3.59}$$

$$HS^- + H^+ = H_2S(aq) \tag{3.60}$$

$$H_3SiO_4^- + H^+ = H_4SiO_4(aq) \tag{3.61}$$

$$NH_3 + H^+ = NH_4^+ \tag{3.62}$$

$$CH_3COO^- + H^+ = CH_3COOH^\circ \tag{3.63}$$

The reaction (3.63) forming acetic acid is representative of a number of similar reactions for the short-chain aliphatic acids, which are commonly found in formation waters, and which can seriously affect the alkalinity titration. Willey *et al.* (1975) report cases where these organic anions account for up to 100% of the measured alkalinity. Unless corrected for these effects, the titration yields a "total alkalinity", rather than a "carbonate alkalinity", no matter how it is reported.

A Terminology Problem There are theoretically several different alkalinity end-points, each with its own name, and there is also some confusion in the usage of "carbonate alkalinity" and "total alkalinity". In this book we discuss the only end-point in common use (*p*H 4.3).[7] We use "total alkalinity" to refer to the general case in which all ions which affect the titration are included, and "carbonate alkalinity" to refer to any titration or corrected titration in which only carbonate and/or bicarbonate ions are neutralized.

To correct the titration alkalinity for the concentrations of such non-carbonate ions is not a simple task. Ideally, we should know the speciation of each component at both the actual *p*H of the solution and at the *p*H of the titration end-point, although approximate methods may be sufficient (Langmuir, 1997; Stumm and Morgan, 1996). The best way to perform this alkalinity correction is to use a geochemical modeling program such as PHREEQC to perform a titration on the sample composition, using various amounts of trial carbonate contents, to determine the carbonate content which results in the observed amount of acid used. This automatically takes care of all competing factors. An example of this is presented in Chapter 8.

---

[7]Other *p*H values from 3.5 to perhaps 5 may also be used.

Example   Titration Alkalinity

In determining the alkalinity of a water sample having a pH of 7.5, 200 ml of the sample is titrated with 0.02 N H_2SO_4 (0.02 moles H^+ per liter). It is found that 26.1 ml of acid is required to reach the methyl orange end-point. What is the total alkalinity?

$$0.0261 \text{ liters acid} \times 0.02 \text{ moles } H^+ \text{ per liter} \times 1000 \text{ ml}/200 \text{ ml}$$

$$= 0.00261 \text{ moles } H^+ \text{ per liter of sample}$$

The titration alkalinity is then 0.00261 mol L^{-1}, or 2.61 meq L^{-1}.

This quantity of moles of H^+ must be converted to the equivalent amount of $CaCO_3$, recalling that 1 mole of CO_3^{2-} is equivalent to (will neutralize) 2 moles of H^+. The number of grams of $CaCO_3$ equivalent to 0.00261 moles of H^+ is then

$$0.00261 \text{ moles } H^+ \times 100.0894 \text{ grams per mole} \times \tfrac{1}{2}$$

$$= 0.131 \text{ grams } CaCO_3 \text{ per liter of sample}$$

The titration alkalinity is then 0.131 g L^{-1}, or 131 mg L^{-1} "as $CaCO_3$".

To report the alkalinity "as HCO_3^-", we not only substitute 61.0174 for 100.0894 (see Table 3.5), but we omit the $\tfrac{1}{2}$ factor because HCO_3^- neutralizes only one H^+, so

$$0.00261 \text{ moles } H^+ \times 61.0174 \text{ grams per mole}$$

$$= 0.159 \text{ grams } HCO_3^- \text{ per liter of sample}$$

The titration alkalinity is then 0.159 g L^{-1}, or 159 mg L^{-1} "as HCO_3^-".

3.9.4 The Alkalinity to Carbonate Component Correction

Suppose now that we have a carbonate alkalinity measurement – either of an unpolluted fresh water, or a measurement corrected for the presence of other anions. This quantity (defined by Equation (3.57)) is obviously somehow related to the quantity we need, which is $m_{CO_3,total}$ (Equation (3.51)), but there are two problems:

- $m_{CO_3^{2-}}$ is included twice in the alkalinity, but we only need it once;

- the alkalinity does not include $m_{H_2CO_3^*}$.

Essentially, we have three unknowns, $m_{H_2CO_3^*}$, $m_{HCO_3^-}$, and $m_{CO_3^{2-}}$, and three equations relating them, Equations (3.53), (3.54), and (3.57). We do not count m_{OH-} or m_{H+} as unknowns, because in those few cases where they become significant, we can estimate them from the pH value. Thus (at 25°C)

$$a_{H^+} = 10^{-pH} \tag{3.64}$$

and

$$a_{OH^-} = 10^{14-pH} \tag{3.65}$$

and, from Equation (3.16),

$$m_{H^+} = a_{H^+}/\gamma_{H^+} \tag{3.66}$$

and

$$m_{OH^-} = a_{OH^-}/\gamma_{OH^-} \tag{3.67}$$

where γ_{H^+} and γ_{OH^-} can be obtained from a speciation model, or may be estimated based on experience. For example, in a solution with a pH of 12, $a_{OH^-} = 10^{-2}$, and if γ_{OH^-} is 0.8, m_{OH^-} is $0.01/0.8 = 0.0125$. This number could then be used to correct the alkalinity value before solving the equations below.

Solving for the three unknowns, and letting A represent the (possibly corrected) alkalinity, we find they are

$$m_{H_2CO_3^*} = \frac{a_{H^+}^2 \gamma_{HCO_3^-} \gamma_{CO_3^{2-}} \cdot A}{\left[K_{3.53} \cdot \gamma_{H_2CO_3^*} \left(2\, K_{3.54} \cdot \gamma_{HCO_3^-} + a_{H^+} \gamma_{CO_3^{2-}} \right) \right]} \tag{3.68}$$

$$m_{HCO_3^-} = \frac{A \cdot a_{H^+} \cdot \gamma_{CO_3^{2-}}}{(2\, K_{3.54} \cdot \gamma_{HCO_3^-}) + (a_{H^+} \cdot \gamma_{CO_3^{2-}})} \tag{3.69}$$

$$m_{CO_3^{2-}} = \frac{A \cdot K_{3.54} \cdot \gamma_{HCO_3^-}}{(2\, K_{3.54} \cdot \gamma_{HCO_3^-}) + (a_{H^+} \cdot \gamma_{CO_3^{2-}})} \tag{3.70}$$

Again, we need estimates of the three activity coefficients, but they can be obtained from a preliminary speciation model, or from experience. The equations are presented in their complete form for reference, but in practice they can be greatly simplified. For example, in solutions having a pH less than about 8.5, Equation (3.70) can be ignored, and the other two simplify to

$$m_{H_2CO_3^*} = \frac{a_{H^+} \cdot A \cdot \gamma_{HCO_3^-}}{K_{3.53} \cdot \gamma_{H_2CO_3^*}} \tag{3.71}$$

Remember that, after calculating $m_{H_2CO_3^*}$, the speciation model should be recalculated if $m_{H_2CO_3^*}$ contributes significantly to $m_{CO_3,total}$.

Example   Calculation of $m_{CO_3,total}$

Consider again our titration example (p. 63). If we had only the alkalinity reported "as $CaCO_3$", we would have to convert this back to milliequivalents per liter or moles per liter by reversing the calculation:

$$131 \text{ mg L}^{-1} \text{ "as } CaCO_3\text{"} = 0.131/100.0894 \text{ moles } CaCO_3$$

$$= \frac{0.131}{100.0894} \times 2 \text{ moles } H^+$$

$$= 0.00261 \text{ moles } H^+$$

So $A = 0.00261 \text{ mol L}^{-1}$. From a speciation model, or from experience, we know that $\gamma_{H_2CO_3^*} = 1$, $\gamma_{HCO_3^-} = 0.8$, and $\gamma_{CO_3^{2-}} = 0.4$, and we know that $K_{3.53} = 10^{-6.37}$ and $K_{3.54} = 10^{-10.33}$.

Inserting all these numbers into Equations (3.68), (3.69), and (3.70), we find

$$m_{H_2CO_3^*} = 0.00015$$

$$m_{HCO_3^-} = 0.00259$$

$$m_{CO_3^{2-}} = 7.68 \times 10^{-6}$$

As expected from the pH value of 7.5, the $H_2CO_3^*$ content is relatively minor compared to $m_{HCO_3^-}$, and $m_{CO_3^{2-}}$ is quite insignificant. The total carbonate content $m_{CO_3,total}$ is in this case $0.00015 + 0.00259 + 7.68 \times 10^{-6} = 0.00276$ molal, not much different from the alkalinity result alone. Needless to say, though, lower pH values will result in much more significant $H_2CO_3^*$ contributions. For example, had the pH been 6.5, $m_{H_2CO_3^*}$ would be 0.00155, and the total carbonate $m_{CO_3,total}$ would be 0.00416, almost twice the alkalinity value. Virtually the same $m_{H_2CO_3^*}$ is obtained by using Equation (3.71) in this case, but for pH ≥ 8.5 it would be inappropriate.

3.10   Acidity

Evidently, all this discussion about alkalinity is only useful for solutions having a pH greater than about 4.3. But many water samples in the environmental field are highly acid. How is the titration method used to determine the carbonate content of a solution with, say, a pH of 3.0?

3.10.1   Titration Acidity

The acidity titration is essentially the reverse of the alkalinity titration. The solution is titrated with a strong base, such as NaOH, to an end-point which is commonly a fixed pH of 8.3, as determined by an indicator color change or pH electrode.

"Carbonate Acidity"

If carbonate is the only important acid present, the titration results in the reactions

$$H_2CO_3^* + OH^- = HCO_3^- + H_2O \tag{3.72}$$

and

$$HCO_3^- + OH^- = CO_3^{2-} + H_2O \tag{3.73}$$

and the result could be termed a "carbonate acidity".

"Non-carbonate acidity" and mineral acidity

However, many other acids may also be present, using up the titrated OH^- in reactions such as

$$HSO_4^- + OH^- = SO_4^{2-} + H_2O \tag{3.74}$$

$$H_2S(aq) + OH^- = HS^- + H_2O \tag{3.75}$$

$$Al^{3+} + 4\,OH^- = Al(OH)_4^- \tag{3.76}$$

$$CH_3COOH + OH^- = CH_3COO^- + H_2O \tag{3.77}$$

and so on. If appreciable quantities of species such as these are present, the result of the acidity titration can be termed a "total acidity", analogous to our discussion of total alkalinity.

Another Terminology Problem   Consider for a moment that $H_2CO_3^*$ is the only significant acid in our solution – we have carbonate acidity. A titration to pH 8.3 converts all $H_2CO_3^*$ to HCO_3^- by reaction (3.72), and the acidity is a measure of $m_{H_2CO_3^*}$ (or strictly speaking of $m_{H_2CO_3^*} + m_{H^+}$ for extremely acid solutions). However, readers of standard references such as Stumm and Morgan (1996) will find that "acidity" is defined in a somewhat different way. Actually, it is just a question of what end-point is used for the titration. Acidity titrations can be carried out all the way to pH 11, converting carbonate in all forms to CO_3^{2-}, and this is often termed the "acidity". However it is now normal analytical practice to stop the titration at pH 8.3. Acidity determined in this way is called "CO_2-acidity" by Stumm and Morgan (1996).

3.10.2   The Acidity to Carbonate Component Correction

Titrating an acid solution to a pH of 8.3 converts all the $H_2CO_3^*$ species to HCO_3^- by reaction (3.72). In §3.9.4 we saw that for solutions having a pH less than 8.5 (which would include most natural solutions), the alkalinity titration is essentially a measure of $m_{HCO_3^-}$. Therefore for most natural solutions, in which other reactions such as (3.58)–(3.63) or (3.74)–(3.77) are not significant, the sum of the alkalinity and acidity titrations is a measure of the total carbonate content.[8]

---

[8]If the acidity titration is carried out to pH 11, it is *twice* the total carbonate.

Correcting the titration acidity for the presence of competing acids in order to obtain a better estimate of $m_{H_2CO_3^*}$ presents the same problems as discussed previously (§3.9.3), and the same remedy is proposed, i.e., the solution analysis should be "titrated" using a program such as PHREEQC, as discussed in Chapter 8.

3.10.3   Alkalinity and Acidity: A Summary

- Geochemical modeling programs need to have information about the total carbonate concentration (Equation (3.51)).

- Chemical analyses often provide this information in the form of alkalinity and/or acidity titration results.

- An alkalinity titration is the quantity of acid (per liter of solution) required to lower the pH to about 4.3. An acidity titration is the quantity of base (per liter of solution) required to raise the pH to about 8.3. These quantities may be reported as "$meq\,L^{-1}$", "$mg\,L^{-1}$ as calcite", or "$mg\,L^{-1}$ as HCO_3^-".

- In general, field measured alkalinity data are preferred over laboratory measured data when both are available.

- In unpolluted fresh waters, the acidity in milliequivalents per liter is approximately equal to $m_{H_2CO_3^*}$ in millimoles per kilogram, and the alkalinity in milliequivalents per liter is approximately equal to $m_{HCO_3^-}$ in millimoles per kilogram. The sum is the total carbonate concentration.

- Some programs (e.g., MINTEQA2) will accept an alkalinity value, and will calculate the total carbonate from this. Others (e.g., EQ3) will not accept alkalinities, so the conversions must be performed before using the program.

- Many natural waters, including most polluted waters, contain acids and bases other than the carbonate species, so that the alkalinity and acidity titrations are not directly related to the carbonate contents. In these cases, approximate corrections can be made by hand, but it is preferable to use a computerized titration to make the correction, as described in Chapter 8.

- In very dilute, very acid, or very basic solutions, corrections for m_{H^+} or m_{OH^-} may be required.

3.11   The Local Equilibrium Assumption

In §3.1.2 we pointed out that thermodynamically based geochemical models can in principle only be used successfully for natural systems which exhibit areas of local equilibrium. We must now examine this idea more closely, and develop criteria for deciding whether or not the local equilibrium assumption (LEA) is a valid approximation in given systems. The conditions under which the LEA is applicable to groundwater and geological systems have been discussed extensively (Bahr, 1990; Bahr and Rubin,

Example   Titration Acidity and Alkalinity

250 ml of a sample of well water, pH 6.6, requires addition of 5.8 ml of 0.1 N NaOH to raise its pH to 8.3. The same volume of water requires the addition of 12.1 ml of 0.1 N HCl to lower its pH to 4.3. The pH is less than 8.5, so we can forget about $m_{CO_3^{2-}}$.

- **Carbonate Alkalinity**

$$0.0121 \text{ liters acid} \times 0.10 \text{ moles H}^+ \text{ per liter} \times 1000 \text{ ml}/250 \text{ ml}$$
$$= 0.00484 \text{ moles H}^+ \text{ per liter of sample}$$

$$\text{carbonate alkalinity} = 4.84 \text{ meq L}^{-1}$$
$$= 242 \text{ mg L}^{-1} \text{ as CaCO}_3$$
$$= 295 \text{ mg L}^{-1} \text{ as HCO}_3^-$$

- **Carbonate Acidity**

$$0.0058 \text{ liters base} \times 0.10 \text{ moles OH}^- \text{ per liter} \times 1000 \text{ ml}/250 \text{ ml}$$
$$= 0.00232 \text{ moles OH}^- \text{ per liter of sample}$$

$$\text{Carbonate Acidity} = 2.32 \text{ meq L}^{-1}$$
$$= 116 \text{ mg L}^{-1} \text{ as CaCO}_3$$
$$= 142 \text{ mg L}^{-1} \text{ as HCO}_3^-$$

- **Total Carbonate**

$$\text{carbonate alkalinity} \approx m_{HCO_3^-}$$
$$\text{so} \quad m_{HCO_3^-} \approx 0.00484$$
$$\text{Carbonate Acidity} \approx m_{H_2CO_3^*}$$
$$\text{so} \quad m_{H_2CO_3^*} \approx 0.00232$$
$$\text{then} \quad m_{CO_3,total} \text{ (as in Equation (3.51))} \approx 0.00484 + 0.00232$$
$$= 0.00716$$

This number (0.00716) would be entered as the carbonate basis species in programs which do not accept an alkalinity.

Alternatively, we could enter the pH and the carbonate alkalinity (0.00232 meq L^{-1}) into Equations (3.68)–(3.70). Using the same activity coefficients as before, we get $m_{H_2CO_3^*} = 0.00228$, $m_{HCO_3^-} = 0.00484$, and $m_{CO_3^{2-}} = 1.80 \times 10^{-6}$, giving $m_{CO_3,total} = 0.00712$. The slight differences are due to the various assumptions we have made.

1987; Knapp, 1989; Lichtner, 1991, 1993; Phillips, 1991; Raffensperger and Garven, 1995; Sanford and Konikow, 1989; Valocchi, 1985).

The following discussion is taken directly from Knapp (1989), to which the reader is referred for more details.

3.11.1 Scales of Interest

The question of fluid–solid phase equilibrium arises not only in environmental problems, but in studies of diagenesis, long range flow in sedimentary basins, ore genesis, magmatic–hydrothermal systems, regional metamorphism, and laboratory experimental systems. In each of these real systems, local equilibrium (LEQ) in theory requires that any disequilibrium condition relax instantaneously to an equilibrium state. In reality, this relaxation occurs over a finite time (t) and, for a fluid-flow system, a finite distance (l). Knapp (1989) points out that each of these types of systems has a characteristic *scale of interest*, which is hundreds of meters or kilometers in studies of sedimentary basins, but perhaps microns in studies of surface processes. If the problem is defined on the kilometer scale, then disequilibrium over distances of centimeters is insignificant. The problem then is to determine, for a given system, the time required for a system in disequilibrium to reach equilibrium (t_{eq}), and the distance the fluid has moved in that time period (l_{eq}).

In geochemical models, these quantities represent the smallest time period for incremental steps in a simulated titration, or the smallest distance between grid points in a finite element or finite difference grid, if LEQ is to be a valid assumption. Or, as Knapp puts it, reactive transport calculations assuming LEQ are good approximations only if t_{eq} is less than the size of the time step, and l_{eq} is less than the distance between adjacent grid points.

3.11.2 Calculation of t_{eq} and l_{eq}

This problem is obviously enormously complex, with many different factors involved, especially if we wish to consider the range of conditions mentioned above. Knapp (1989) simplifies the problem by (i) collapsing many of the controlling parameters into two dimensionless parameters; and (ii) considering only one spatial dimension, one heterogeneous reaction, and one component.

Consider a heterogeneous system at equilibrium, and what happens when that equilibrium is instantaneously disturbed. The system is water flowing through a quartz sandstone aquifer, with the water everywhere in equilibrium with the quartz, at some fixed temperature (obviously we are considering chemical equilibrium, not hydraulic equilibrium). At one point along the flow path, we introduce a pulse of pure water, i.e., water greatly undersaturated with quartz. As the fluid moves along, this pulse of fluid gradually dissolves quartz, and at the same time is subject to flow, diffusion, and dispersion. This situation is pictured in Figure 3.3.

The various factors controlling the time it takes for this pulse of fluid to reach equilibrium, and the distance it travels during that time, are combined into the Damköhler

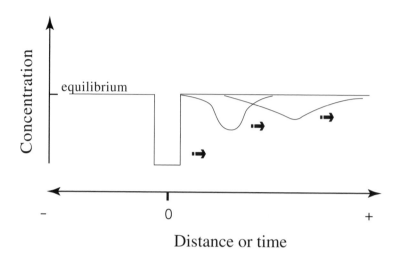

Figure 3.3. The concentration of silica in a fluid along a one-dimensional flow path in a quartz sandstone, initially at equilibrium. A pulse of dilute fluid is introduced at time 0 or distance 0, and gradually relaxes towards equilibrium. Reaction toward equilibrium is driven by irreversible dissolution of quartz, diffusion, and dispersion. The small arrows indicate the flow direction (after Knapp, 1989, Figure 1).

number (Da) and the Peclet number (Pe),

$$Da = \frac{v \overline{S} a_{H^+}^n L}{c_{eq} \overline{v}}$$ (3.78)

and

$$Pe = \frac{\overline{v} L}{D_{eff} + D}$$ (3.79)

where

a_{H^+}	is	hydrogen ion activity
n	is	the exponent for the hydrogen ion activity for the specified reaction
v	is	the stoichiometry of the component in the phase considered (e.g., $v = 1$ for SiO_2 in quartz)
c_{eq}	is	the equilibrium concentration
L	is	an arbitrarily selected characteristic length
$\overline{S}$	is	specific surface area (m^2 of mineral per m^3 of fluid)
$\overline{v}$	is	true fluid velocity
D_{eff}	is	the effective diffusion coefficient
D	is	the dispersion coefficient

Barred quantities are averages over a representative volume.

The Damköhler number expresses the rate of reaction relative to the advection or fluid flow rate. A large Da value means that reaction is fast relative to transport and that aqueous concentrations may change rapidly in time and space; the temporal and spatial scale of LEQ may be relatively small. The Peclet number expresses the importance of advection relative to dispersion in transporting aqueous compounds. A large Pe value means that advection dominates, which may result in large concentration gradients; a small Pe value suggests that dispersion dominates, which promotes mixing in the fluid phase.

In addition, we define a dimensionless time (τ) and a dimensionless distance (X),

$$\tau = \frac{t\bar{v}}{L} \tag{3.80}$$

and

$$X = \frac{x}{L} \tag{3.81}$$

where x is distance.

Having set up the initial and boundary conditions, the relevant partial differential equations are solved, resulting in

$$2\,\mathrm{Da}\,\tau_{\mathrm{eq}} + \ln(\tau_{\mathrm{eq}}) - \ln(\mathrm{Pe}) - 6.679316 = 0 \tag{3.82}$$

where τ_{eq} is the dimensionless time required for an initial dissolution impulse to relax to 99% of c_{eq}, the equilibrium concentration. The dimensionless distance, X_{eq}, required for relaxation to 99% of c_{eq} is equal to τ_{eq}. Equations (3.80) and (3.81) are used to transform τ_{eq} into the more useful variables t_{eq} and l_{eq}, thus

$$l_{\mathrm{eq}} = \tau_{\mathrm{eq}} L \tag{3.83}$$

and

$$t_{\mathrm{eq}} = \frac{\tau_{\mathrm{eq}} L}{v} \tag{3.84}$$

Equation (3.82) thus relates the time required to reach equilibrium and the distance traveled during that time to virtually all the other factors controlling the attainment of heterogeneous equilibrium, at least for one reaction and in one dimension. It can be solved for τ_{eq}, given various values of Da and Pe, resulting in Figure 3.4.

The contours of $\log(\tau_{\mathrm{eq}})$ describe a hyperbolic surface, shown in three dimensions in Figure 3.5. Thus there is a region where the time and distance to equilibrium is dependent only on Da (reaction dominated), and there is another region where they are independent of Da (transport or advection dominated). Local equilibrium can occur in both domains. Most natural environments with elevated temperatures fall in the reaction dominated domain, where the effects of dispersion and diffusion can safely be ignored.

Transport domination of τ_{eq} and X_{eq} occurs for small values of Da and Pe. This occurs where reaction rates are slow relative to advection rates, and at large scales of interest. This occurs at low temperatures and relatively rapid fluid flow, perhaps typical of some diagenetic and sedimentary basin studies. Under these conditions, the uncertainties in rate constants and reactive surface areas are not significant in determining the temporal and spatial scale of LEQ. An estimate of conditions for sedimentary basins by Raffensberger and Garven (1995) is shown in the left-hand shaded area in Figure 3.4.

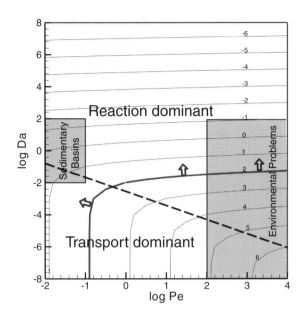

Figure 3.4. Contours of $\log(\tau_{eq})$, which is equal to $\log(X_{eq})$, for a range of Damköhler and Peclet numbers. The dashed line separates a region where τ_{eq} and X_{eq} depend only on Da (reaction dominated) from a region where they depend only on Pe (transport or advection dominated). The shaded areas are discussed in the text. The arrows point to the region where the modeling grid size can be less than 100 meters.

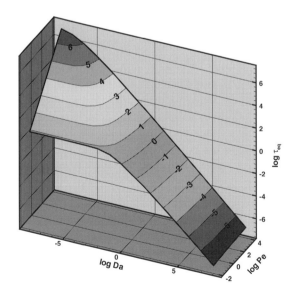

Figure 3.5. The $\log(\tau_{eq})$ surface from Figure 3.4 shown in three dimensions.

t_{eq} and l_{eq} in Environmental Situations

Unfortunately, LEQ would appear to be a questionable approximation in what Knapp (1989) refers to as "human controlled environments" due to characteristically large fluid velocities and low temperatures. Using the values of Da and Pe suggested by Knapp, it appears that environmental problems lie in the small Da, large Pe region, as indicated by the right-hand shaded area in Figure 3.4. If we were to assume that a finite difference or finite element grid should be no larger than 100 meters, we are then restricted to quite a small area, as indicated by the arrows.

This analysis by Knapp (1989) is useful in defining and clarifying the LEQ problem in a quantitative way. Unfortunately, most of the parameters required to define Da and Pe in real environmental situations at the present time are poorly known. This is especially true of $\overline{S}$, the ratio of solid phase surface area to the volume of fluid reacting with the solid. The shaded areas, and even the positions of the contours, are of questionable significance in any practical sense. What we learn from this exercise is that the low temperature, high fluid flow conditions typical of environmental problems mean that the times and distances required to attain equilibrium may be quite large, and that we should be aware that this will be one of the limiting factors in applying our thermodynamically based models. Unfortunately, we will not often be able to say exactly what times or distances are required in any given case. This may seem unsatisfactory, but it is only one of several such limiting factors we will be discussing, and it emphasizes again that despite the mathematical exactness of the programs we use, the conclusions we draw from the results are always a result of our own judgment.

The fact that the local equilibrium assumption is highly questionable highlights the importance of kinetics (Chapter 11) and may appear to invalidate much of geochemical modeling. This is not the case. In our view, modeling is always valuable, whatever the state of our knowledge. Refer to §1.4.2, page 16, for a more detailed discussion of this point.

3.12   Summary

This chapter has outlined the basic knowledge required to perform geochemical modeling. The transformation of Gibbs energies into equilibrium constants, the use of activities and activity coefficients, and the relationship of these to concentrations are fundamental. A clear distinction between components and species is required in order to input data to programs correctly, and to interpret the results. Programs need to have completely specified systems, and the Phase Rule can be used to determine the number of pieces of data (degrees of freedom) required to do that. All of these concepts are explained in more detail in many standard sources.

The implementations of the alkalinity and acidity concepts can be quite confusing, but analyses are often reported in these terms, and they must then be used to determine the carbonate component. Doing this sometimes involves using a titration program. We have also emphasized that most of our modeling programs assume a state of complete equilibrium, which places constraints on the times and distances we can consider in our models. Unfortunately, at the present time these constraints are poorly defined.

4

Computer Programs for Geochemical Modeling

4.1 Codes, Databases, and Models

The most widely used geochemical modeling programs consist of a computer code plus a related file of data called a database. The database contains thermodynamic and kinetic parameters. The code uses the thermodynamic and kinetic parameters in the database and concentrations or other constraints as input, and produces results that describe a geochemical model for a particular chemical system.

In this chapter we shall limit our discussion of the codes to speciation–solubility and reaction path modeling codes. Coupled reactive mass transport codes are much more mathematically and computationally complex. The readers can find recent discussions in Lichtner (1996) and Steefel and MacQuarrie (1996).

4.1.1 The Code

A computer modeling code or program is a set of computer commands that include algorithms to solve a set of mathematical equations describing chemical equilibria,

Algorithms	The numerical techniques embodied in the computer code
Computer code	The assembly of numerical techniques, bookkeeping, and control language that represents the model from acceptance of input data and instruction to delivery of output
Model	The assembly of concepts in the form of mathematical equations that portrays understanding of a natural system

Table 4.1. Definitions of terms commonly used in modeling by ASTM (1984).

> The following are required to perform geochemical modeling:
>
> - a computer code,
>
> - a thermodynamic, kinetic, and surface property database,
>
> - chemical and physical measurements of the system of concern.

chemical mass balance, various kinds of bookkeeping, and input and output controls (see Table 4.1). The code itself only deals with the mathematical framework and computation tasks, with no reference to any particular chemical system. It is designed this way so as to be general. As a result, the mathematical framework for all such codes is similar. They all essentially try to solve the same set of algebraic equations, although the mass balance accounting and computational techniques for solving these equations may vary in each code. These variations mean that codes have different abilities in terms of convergence and handling of special and extreme cases, but generally they do not result in significantly different answers, given an identical geochemical problem (Nordstrom *et al.*, 1979).

The codes published by government agencies (MINTEQA2, EQ3/6, and PHREEQC) have all gone through some, at times lengthy, quality assurance/quality control (QA/QC) procedures, and are verified against other codes or hand calculations for their mathematical functionality. In other words, they perform the calculations they are supposed to do (see Chapter 2).

The mathematical formulation of equilibrium speciation–solubility models and the numerical methods to solve the algebraic equations are described elsewhere, and we shall not repeat them here. The reader can consult Anderson and Crerar (1993), Bethke (1996), and Westall *et al.* (1976). Mathematical formulations for the reaction path models are described in detail in Anderson and Crerar (1993) and Bethke (1996).

4.1.2   The Database

The thermodynamic databases that accompany most modeling codes were prepared intentionally to be separate from the codes. This means equilibrium constants are not "hard-wired" into the codes, which makes it much easier for users to change the values of the equilibrium constants in the database, or add/delete a reaction from the database without affecting the functionality of the codes. Many codes (e.g., PHREEQC and EQ3/6) allow users to modify the equilibrium constants in the input file as well as in the database itself (see the Appendix).

None of the thermodynamic databases accompanying the modeling codes are comprehensive compilations of all the aqueous and mineral species we may encounter. Neither has it been ensured that the thermodynamic data are internally consistent. Code developers and releasing agencies make no statement about what are the "best" thermodynamic properties of a solid or aqueous species or about the internal consistency of the data; they assume this to be the responsibility of the modeler.

4.2   Review of Popular Computer Programs

There are now dozens of readily available computer programs which can perform the calculations central to geochemical modeling. Of course, these codes have widely varying capabilities (Table 4.2). A summary of most existing codes is presented by Mangold and Tsang (1991), by Paschke and van der Heijde (1996), and by van der Heijde and Elnawawy (1993). We will concentrate on a few that are most commonly used in this field. Some are included on the accompanying CD-ROM.

- MINTEQA2 Version 3.11 (Allison *et al.*, 1991). A program sponsored by the US Environmental Protection Agency. Its primary purpose is speciation modeling, including redox, ion-exchange, and several surface complexation models. Dzombak and Morel's (1990) surface complexation constants are included in a separate database file.

  The program is available free of charge from US EPA by contacting:

  Model Distribution Coordinator, Center for Exposure Assessment Modeling, 960 College Station Road Athens, GA 30605-2700, USA.

  E-mail: ceam@epamail.epa.gov; telephone: (706)-355-8400.

- PHREEQC (Parkhurst and Appello, 1999). Produced by the US Geological Survey. A speciation and reaction path program. PHREEQC uses the *C* language and is capable of performing speciation and solubility, reaction path, inverse mass balance modeling, and one-dimensional advective–dispersive–reactive transport calculations. The inverse mass balance modeling module has the capability of taking analytical uncertainties into account. MINTEQA2 and PHREEQC are the two most widely used programs today.

  The program can be downloaded free of charge from the US Geological Survey web site:

  http://wwwbrr.cr.usgs.gov/projects/GWC_coupled/.

- EQ3/6 (Wolery, 1992). EQ3 is a speciation program, which can be used alone or to prepare input for EQ6, a reaction path program. A kinetics option is also available. The program has a comprehensive database, primarily through the effort of Dr James Johnson. Unlike many other databases, the sources of thermodynamic data are documented in this database.

  The program is available for a nominal fee from:

  Technology Transfer Initiative Program, L-795, Lawrence Livermore National Laboratory, P.O. Box 808, Livermore, CA 94550, USA.
  Telephone: (925)-422-7678.

- NETPATH (Plummer *et al.*, 1994). Produced by the US Geological Survey. An inverse mass balance modeling program for the interpretation of net geochemical mass balance reactions along a hydrologic flow path. This program is capable of modeling ^{14}C, ^{18}O, D, and ^{34}S fractionation.

The program can be downloaded free of charge from the US Geological Survey web site:

http://wwwbrr.cr.usgs.gov/projects/GWC_coupled/.

- SUPCRT92 (Johnson *et al.*, 1992). A different kind of modeling program, SUPCRT92 models thermodynamic data using the HKF model. It provides standard state thermodynamic data and equilibrium constants for a wide range of minerals, gases, and aqueous organic and inorganic species over a very wide range of temperatures and pressures, which are widely used in other modeling programs.

  The program is available free of charge from Dr James Johnson at Lawrence Livermore National Laboratory (e-mail: jwjohnson@llnl.gov).

- EQBRM (Anderson and Crerar, 1993, App. E) This program will do a speciation calculation, using Davies activity coefficients. However, both the species to be solved for, and the equilibrium constants required must be supplied. It is very small and fast, and best suited for conceptual problems rather than real-world problems.

  The source code is listed in App. E of Anderson and Crerar (1993). The source code and excutable file are included in the accompanying CD-ROM, and can be downloaded from

  http://www.geology.utoronto.ca/faculty/anderson.html or

  http://www.pitt.edu/~czhu/book.html.

In addition to these programs, which are freely available and very widely used, there are a few others we will mention from time to time, which are also useful. The first two are available at relatively nominal cost from their authors. The Geochemist's Workbench™ group of programs are available on a commercial basis.

- SOLMINEQ.88 pc/shell (Perkins, 1992), GAMSPATH.99. SOLMINEQ.88 is a speciation program with several other options, developed primarily for petroleum industry applications. GAMSPATH.99 is a general reaction path code developed at the Alberta Research Council. It uses kinetic rates of reaction. Rate laws and constants are specified for each phase in the general database. Both reactant and product phases may be out of equilibrium with the aqueous phase. Both dissolution and precipitation kinetics are considered, including nucleation.

  Inquiries should be directed to Dr Ernest Perkins at

  perkins@arc.ab.ca

  or at the web site of Geochemical Applications & Modelling Software Ltd:

  http://www.telusplanet.net/public/geogams/index.html.

- SOLVEQ, CHILLER (Reed, 1982). SOLVEQ is a speciation program that also prepares input to CHILLER, which is a reaction path program capable of boiling simulations. Inquiries should be directed to Prof Mark H. Reed at

  mhreed@oregon.uoregon.edu

Functions		MINTEQA2	PHREEQC	EQ3/6	GWB
Types of rxn	aqueous speciation	✓	✓	✓	✓
	pptn/dissolution	✓	✓	✓	✓
	solid solutions			✓	✓
	ion-exchange	✓	✓		
	surface complexation				
	constant capacitance	✓			
	double layer	✓	✓		✓
	triple layer	✓			
Act. coefficient	Davis	✓	✓	✓	✓
Models	B-dot		✓	✓	✓
	Pitzer			✓	✓
Redox	disequilibrium		✓		✓
Gas phase		✓	✓	✓	✓
Reaction path	mixing/titration		✓	✓	✓
models	flow through		✓	✓	✓
	flush		✓		✓
	dump option		✓		✓
	fixed activity/fugacity		✓	✓	✓
	slide activity/fugacity		✓		✓
	change of temperature		✓	✓	✓

Table 4.2. Functions of popular geochemical modeling programs.

- MINEQL+ Version 4.0. A Windows-based sibling to MINTEQA2. Although it shares the same lineage with MINTEQA2 to MINEQL (Westall *et al.*, 1976), the Windows interface makes composing the model more transparent to the modeler and viewing modeling results easier, and there are many nice enhancements. The thermodynamic data in the model are mainly from MINTEQA2, which in turn came from WATEQ. Constants in the original MINEQL database that are not in the MINTEQA2 database are also included. This program is widely used by aquatic chemists and engineers. Inquiries should be directed to:

Environmental Research Software, telephone: (207)-622-3340;
e-mail: ersofwr@agate.net.

- The Geochemist's Workbench™. A collection of five geochemical programs by C.M. Bethke and his group at the University of Illinois, capable of performing virtually all options we discuss (except coupled reactive transport), with graphical output (Bethke, 1994, 1996). Inquiries should be directed to

www.rockware.com.

We will say very little about the numerical methods these programs use. An introduction to this subject is provided by Bethke (1996), Smith and Missen (1982), and Van Zeggeren and Storey (1970).

4.3 Databases

All of these programs (except EQBRM) have an associated database which contains the information on species properties and equilibrium constants required by the program. The program uses the data in the database, which may or may not be the best available at the present time. We discuss the question of the quality of the data in §4.3.2.

4.3.1 A Typical Database

The databases associated with geochemical modeling programs consist of a list of the basis species, and a list of secondary or auxiliary species, minerals, and gases, each described in terms of the basis species, and with the equilibrium constant of the reaction linking the secondary species or mineral to the basis species.

For example, the database for the The Geochemist's Workbench™ programs looks like the following. We choose this one because it is relatively readable. Databases for other programs contain essentially the same kind of information, but usually in a more condensed format which is harder for humans to understand.

```
    47 basis species

H2O
      charge=  0.0       ion size=  0.0 A       mole wt.=   18.0152
      2 elements in species
        1.000 O                 2.000 H
Ag+
      charge=  1.0       ion size=  2.5 A       mole wt.=  107.8680
      1 elements in species
        1.000 Ag
Al+++
      charge=  3.0       ion size=  9.0 A       mole wt.=   26.9815
      1 elements in species
        1.000 Al
etc.........................................................

551 aqueous species

AgCl
      charge=  0.0       ion size=  4.0 A       mole wt.=  143.3210 g
      2 species in reaction
        1.000 Ag+              1.000 Cl-
        -3.5100    -3.3100    -3.1300    -2.9900
        -2.9200    -2.9000    -3.0000    -3.4000
AgCl2-
      charge= -1.0       ion size=  4.0 A       mole wt.=  178.7740 g
      2 species in reaction
        1.000 Ag+              2.000 Cl-
        -5.5700    -5.2500    -4.9400    -4.7100
        -4.5700    -4.6000    -4.7000    -5.2000
AgCl3--
```

```
    charge= -2.0        ion size=   4.0 A        mole wt.=   214.2270 g
    2 species in reaction
       1.000 Ag+                   3.000 Cl-
         -5.5100    -5.2500    -5.0300    -4.8800
         -4.8500    -5.0000    -5.3000    -6.0000

etc.................................................................

624 minerals

Alabandite                           type= sulfide
       formula= MnS
       mole vol.=   21.460 cc      mole wt.=   86.9980 g
       3 species in reaction
       -1.000 H+                   1.000 Mn++              1.000 HS-
          0.0101    -0.3811    -0.9405    -1.5286
         -2.2051    -2.8755    -3.6341    -4.7049
Albite                               type= feldspar
       formula= NaAlSi3O8
       mole vol.=  100.250 cc      mole wt.=  262.2230 g
       5 species in reaction
       2.000 H2O                   1.000 Na+               1.000 Al+++
       3.000 SiO2(aq)             -4.000 H+
          3.9160     3.0973     1.9915     0.9454
         -0.0499    -0.8183    -1.5319    -2.5197
etc.................................................................
```

From this, we can see that secondary species $AgCl_2^-$ is related ("linked") to the basis species Ag^+ by[1]

$$AgCl_2^- = Ag^+ + 2\,Cl^-$$

for which, at 25°C,

$$\log K = -5.25$$

and that the mineral albite is related to several basis species by

$$NaAlSi_3O_8 = 2\,H_2O + Na^+ + Al^{3+} + 3\,SiO_2(aq) - 4\,H^+$$

for which, at 25°C,

$$\log K = 3.0973$$

---

[1] Note that the two rows of four numbers in each set of data for secondary species and minerals are the log K values for the reactions at temperatures of 0, 25, 60, 100, 150, 200, 250, and 300°C.

The practical importance of looking into the database in this way is that the equations relating the secondary species to the corresponding basis species show which basis species can be "swapped" (see §5.8.1). Thus, $AgCl_2^-$ can be swapped for either Ag^+ or Cl^-, and $NaAlSi_3O_8$ can be swapped for Na^+, Al^{3+}, $SiO_2(aq)$, or H^+. Also, it is often important to see exactly how the data are labeled. For example, the data for sodic feldspar follow the label "Albite", not "albite", or "Na-feldspar".

4.3.2 Data Quality

The quality of thermodynamic data of all types and in all fields of science is a subject of endless debate. It is also a vexing subject for users of geochemical modeling programs, because, although it is clearly the user's responsibility to ensure the quality of the data used in the program (users are free to edit their database in any way they see fit, no matter where it originated), judging the quality of thermodynamic data is a job for specialists, and even they very often do not agree among themselves. Engi (1992) tells us that in fact

> ... there is as yet no generally accepted way of characterizing adequately the quality of refined thermodynamic data.

A useful discussion of data accuracy is in Nordstrom and Munoz (1986).

No one who creates a database and makes it available for general use will make any guarantee about the accuracy of the data, for obvious reasons, so the user is left in a minefield of conflicting data, because the commonly available databases are certainly not identical.

On the other hand, there is a quite reasonable level of agreement among commonly available databases for the more commonly occurring minerals and aqueous species (Nordstrom *et al.*, 1979). The modeler will likely obtain similar results from various programs if the problems involve only common phases and species. We give an example of this in Chapter 7. It must be emphasized that, although this level of agreement is comforting, it does not by itself mean that the results are accurate or true. There is not only the possibility that the model may have been set up incorrectly, but, more fundamentally, even commonly accepted data may in fact be wrong. This has to do with the notions of precision and accuracy.

Precision and Accuracy

"Accuracy" refers to the degree of agreement between a measured quantity and the true value. "Precision" refers to the reproducibility of that measurement, i.e., some measure of the dispersion of the measurements, the mean of which is reported as the measured value. It is well recognized that high precision is no guarantee of accuracy. If a sample is sent for analysis to ten different analysts, and nine of them agree very closely and the tenth gives quite a different result, it may be the tenth one which is the most accurate, for various possible reasons.

The thermodynamic data for ordered dolomite are commonly agreed upon, and calculations based on the data show that seawater is greatly oversaturated with dolomite, yet it does not precipitate. This is widely accepted. Recently, Lafon *et al.* (1992) have suggested that the dolomite data may be inaccurate, and that dolomite may not in fact be oversaturated. And so the debate continues.

The fact that basic thermodynamic data are imprecisely known means that model results such as pH values, SI values, and so on, will all be to some extent imprecise as well. The imprecision of the input data is propagated through the calculation procedure and appears in the results. The nature of this propagation has not been extensively investigated, but it depends not only on uncertainties in the thermodynamic and analytical data, but also on the nature of the geochemical system involved. See Anderson (1976, 1977) and Criscenti *et al.* (1996) for discussions.

Internal Consistency

Another problem is that of consistency among the data in a database. That is, data for $\Delta_f S°$, $\Delta_f H°$, and $\Delta_f G°$ for compound X may be obtained from different sources, and each may be the best available, but they may not be consistent with $\Delta_f G_X° = \Delta_f H_X° - T \Delta_f S_X°$. Or we may obtain the best possible $\Delta_f G°$ data for each of lime (CaO), calcite, and carbon dioxide, but when put together, we find they are not consistent with equilibrium constant data for the reaction $CaCO_3 = CaO + CO_2$. There are several levels of such "self consistency" that should be satisfied (Engi, 1992, p. 276), and dealing with this problem is another job for specialists. It is unfortunately a very big job, and it has not been carried out to the fullest extent possible for any database commonly used for environmental problems.

What to Do?

This kind of discussion, common among academics, can be discouraging to practioners who have problems to solve. If specialists cannot agree, how should the non-specialists satisfy themselves that their data are accurate? Frankly, it is likely that most do not. There is a kind of "safety in numbers", so that if modelers are using a database that hundreds of others are also using, data errors are at least inconspicuous. Beyond that, the only recourse is to keep abreast of the literature, and to be aware of the occasional updates that are made. On those occasions when it is necessary to locate data for unusual species which are not included in common databases, users are more or less on their own, and, lacking any expertise in locating and evaluating data, the best bet is to find some expert help. A good place to start is the National Institute of Standards and Technology (NIST) of the US Department of Commerce (http://www.nist.gov/srd). The NIST Standard Reference Database 46 is available as a searchable computer program containing almost 83 000 pieces of data on metal–ligand stability constants, with emphasis on metal–organic ligands (Smith and Martell, 1998).

Unfortunately, lack of accuracy and internal consistency in our thermodynamic data may well be the least of our problems. It is easy to imagine that even if we had absolutely accurate data for every aqueous species and mineral, and accurate models for activity coefficients in all phases, predictions based on present models would still be inaccurate. This is the point made by Kraynov (1997). The deeper problem is the inadequacy of the conceptual framework of our geochemical models themselves, including their reliance for the most part on the local equilibrium assumption (§3.11).

Nature is tremendously complex, and our models are relatively simple. Only a gradual increase in our understanding over time will alleviate this problem.

4.4 Chemical Concentration Units

The units most commonly used in reporting analyses of natural waters are milligrams per liter (mg L^{-1}), but virtually all geochemical calculations are made on a weight basis, using moles or millimoles per kilogram of water. In dilute solutions, mg L^{-1} is the same as parts per million (ppm) or mg kg^{-1}. At higher concentrations, however, the solution density should be taken into account according to

$$ \text{mg kg}^{-1} = \frac{\text{mg L}^{-1}}{\text{solution density}} $$

Most programs have a provision for inputting the density of the solution if it is known. If it is not, a value of 1.0 g cm^{-3} is assumed.

The densities of NaCl solutions at 25°C are shown in Table 4.3.

Molality	0.1	1	2	3
Density (g cm^{-3})	1.0012	1.0362	1.0723	1.1058

Table 4.3. The densities of NaCl solutions at 25°C.

Natural solutions are not the same as pure NaCl solutions, but these numbers will provide a rough idea of how much error is involved in ignoring the density correction at various concentrations. For example, we would expect to introduce approximately a 3% error in our results by ignoring the density of a 1 molal solution (about 58 000 mg L^{-1} total dissolved solids, if all NaCl).

Most current programs will provide a choice of units for input data, and will present the results in a variety of units.

4.5 Examples of Input/Output

The geochemical modeling market has not yet been discovered by Microsoft, so geochemical modeling programs have no context-sensitive help screens, no "wizards" for guidance, and the documentation is sometimes less than completely helpful. In other words, inputting our valuable data into one of these programs is often a challenge, until it becomes completely familiar. Interpreting the output is easier, but still requires some knowledge of basic principles (see, e.g., §5.8.1).

4.5.1 Program Input

These days, in the age of personal computers, data input is invariably done by typing the data onto the terminal screen. This may be while the modeling program is actually running, in which case the program prompts the user with questions on the screen, requiring a response, or a separate file may be prepared before running the program, which the program then reads to obtain the data. In this case, the formatting of the data must conform to what the program expects, and this is commonly done either by using a "template" file or by preparing a script having keywords followed by data. Another option is to have a separate data preparation program, which prompts the user interactively for data, then prepares a file which is used by the modeling program.

A *template file* is simply a file in which text and symbols are arranged with delimited areas for data, simplifying the process of figuring out the proper format. It is quite commonly just an old input file, in which the data are entered for a new problem. The essence of the process, however, is that each piece of data must be in a certain place in the file.

A *script* is a text file prepared with an editor, independently of the modeling program.[2] Data can appear anywhere in the file, but must be preceded by a keyword that the program recognizes, so that the program knows what to do with the data which comes after the keyword.

Given the rate of software development, a good way to ensure that a book such as this becomes quickly out of date is to base it closely on existing programs. Nevertheless, we must say something about data input, and the methods just described probably will not change much.

Among the programs listed above, SUPCRT92 uses interactive prompting, SOLVEQ and MINTEQA2 use separate data file preparation programs which use interactive prompting (called GEOCAL in the case of SOLVEQ; PRODEFA2 in the case of MINTEQA2), EQ3/6 uses a template file, and PHREEQC and NETPATH use script input. SOLMINEQ.88 pc/shell has a unique system of menus to be filled in, depending on the options chosen, which is basically a template system.

We obviously cannot go into all the details of data input for each program, but below we show examples of each type of input, with some comments.

Interactive Input

Whether the modeling program itself prompts the user for data (e.g., SUPCRT92), or this is done by a program which prepares a file to be read by the modeling program (e.g., PRODEFA2 for MINTEQA2) makes little difference to the user. The interactive menus for MINTEQA2 are shown below. They appear in four "levels":

```
_____ M A I N   M E N U :   S E L E C T   O P T I O N _____PROB #  1__

  1 = EDIT LEVEL I   (Change ionic strength, pH, Eh, temperature, adsorption
      parameters, number of iterations, precipitation options, etc.)

  2 = EDIT LEVEL II   (Specify components, gas, redox, aqueous, and mineral
      species, adsorption sites and reactions, add new species of all types)

  3 = EDIT LEVEL III   (Check, individually edit all entries)

  4 = EDIT LEVEL IV   (Sweep a range of pH, pE, or dissolved concentration;
```

[2]The Geochemist's Workbench™ is a hybrid, in that the user prepares a script input of keywords and data, but interactively as the program is running, allowing the program to respond with help if something is incorrect. The script can also be saved and modified or used again.

```
         Designate an auxiliary MINTEQA2 output file to receive equilibrated
         output for spreadsheet import.)

   M = MULTI-PROBLEM GENERATOR

   X = EXIT  (Write the current problem to the new MINTEQA2 input file
       and EXIT PROGRAM)

      ENTER CHOICE  >
```

Edit level I is the main data input screen. It looks like this:

```
_____ EDIT LEVEL I _____PROB #  1___
 1 Title 1:
 2 Title 2:
 3 Temperature (Celsius):  25.00
 4 Units of concentration: MOLAL
 5 Ionic strength:  TO BE COMPUTED
 6 Inorganic carbon is not specified.
 7 Terminate if charge imbalance exceeds 30% ? NO
 8 Oversaturated solids ARE NOT ALLOWED to precipitate. EXCEPTIONS: Solids
   listed in this file as TYPE -III (Infinite), -IV (Finite) or -V (Possible).
 9 The maximum number of iterations is:  40
10 The method used to compute activity coefficients is: Davies equation
11 Level of output: INTERMEDIATE
12 The pH is: TO BE COMPUTED
13 The pe and Eh are: UNDEFINED
99 Choose a different file to modify OR return to output filename prompt.

To change any of the above entries or to explore other possible values,
enter the number to the left of the entry.  Press ENTER to accept all settings.

ENTER CHOICE  >
```

Each item has default data already entered, so that only those items that need to
be changed need be chosen. Data is entered by first entering the item number, then
answering further questions. For example, if on entering "6" the question

```
Do you want to specify dissolved inorganic carbon
 in this problem ? (Y,N) >
```

appears and if "y" or "Y" is entered, the following screen results:

```
When alkalinity is specified, no solids are allowed. (Set EDIT LEVEL I Option 8
to zero and specify no TYPE III,IV, or V solids.)
Also, the titration used to determine alkalinity is assumed to be to the
pH that is the equivalence point of the solution).  Otherwise the alkalinity,
factors in the database will not be applicable.

You have the option of specifying alkalinity as a measure of
dissolved inorganic carbon.  Alternatively, you may specify dissolved
inorganic carbon explicitly.  Your choice will generally depend upon the way
carbonate concentration is expressed in the chemical analysis of the
sample you are modeling.

 Do you want to specify alkalinity ? (Y,N) >Y

 Select alkalinity units:
  0 = Return to previous question
  1 = mg/l CO3-2
  2 = mg/l CaCO3
  3 = eq/l
      ENTER CHOICE  >
```

Choosing 1, 2, or 3 at this point allows the user to enter an alkalinity value. MINTEQA2
uses "mg/l as CO_3^{2-}" instead of "mg/l as HCO_3^-", and "eq/l" instead of "meq/l", the
units discussed in §3.9.3.

After Edit Level I, follows Level II, in which the chemical system is set up :

```
_____ EDIT LEVEL II _____PROB #  1_
                     _____ S E L E C T   O P T I O N _____

     1 = Specify AQUEOUS COMPONENTS: TOTAL CONCENTRATIONS or FIXED ACTIVITIES
     2 = Specify AQUEOUS SPECIES not in the database, search the database,
         or alter a database AQUEOUS SPECIES equilibrium constant
     3 = Specify an ADSORPTION MODEL and REACTIONS
     4 = Specify GASES at FIXED partial pressures
     5 = Specify REDOX COUPLES with FIXED activity ratios
     6 = Specify INFINITE SOLID phases
     7 = Specify FINITE SOLID phases
     8 = Specify POSSIBLE SOLID phases
     9 = Specify EXCLUDED SPECIES of any type
     R = RETURN to MAIN MENU

     All choices allow you to browse and return without changing anything;
     Most allow you to search or view a directory of the relevant database.
        ENTER CHOICE ([D] = R) >
```

MINTEQA2 unfortunately uses a rather complex system of index numbers to identify chemical species, minerals, etc., but, as many people can't recall the index numbers or don't want to bother with them, the menu system has a cyclical method of questions and answers which allows the user to choose components, minerals, etc. without knowing the index numbers, or in fact without knowing anything at all about the database. This is an advantage, perhaps, especially for new users, but it is a rather cumbersome method of data entry. It is more difficult to control the model exactly, for example to swap secondary for primary basis species.

With most other programs, we must know exactly how the species are identified in the database, because the program uses our input to search the database. For example, the data for sodic feldspar on p. 80 is entered under the label "Albite". In most programs, if one enters a mass, say, of "albite", the program will search for that label, will not find it, and will report an error.

MINTEQA2 also divides solid phases into three classes, depending on whether they are or are not allowed to precipitate or dissolve completely. See the documentation for more details.

Template File Input

Using a template file to create a data input file is quite simple, but it is still important to understand the meaning of the various categories. EQ3/6 actually has two different data entry formats – an older method requiring exact formatting, and a newer one using the following template, used here to enter the data from sample MW–12 (see Table 6.1):

```
|-----------------------------------------------------------------|
|EQ3NR input file name= MW12.3i                                   |
|Description= "Bear Creek Project, Sample MW12"                   |
|Version level= 7.2                                               |
|Created           Creator=                                       |
|Revised           Revisor=                                       |
|                                                                 |
|  Large amounts of descriptive material can go here.             |
|                                                                 |
|-----------------------------------------------------------------|
|Temperature (C)        | 12.0      |Density(gm/cm3)|  1.00000    |
|-----------------------------------------------------------------|
|Total Dissolved Salts  |           | mg/kg | mg/l  |*not used     |
|---------------------------------------------------------------------|
|Electrical Balancing on |SO4--               | code selects| not performed|
|---------------------------------------------------------------------|
```

```
| SPECIES                | BASIS SWITCH/CONSTRAINT| CONC/ETC  | UNITS OR TYPE |
|-------------------------------------------------------------------------------|
|redox                   |                        | 0.0       |logfO2         |
|H+                      |                        | 6.7       |ph             |
|Na+                     |                        | 265       |mg/l           |
|K+                      |                        | 14        |mg/l           |
|Al+++                   |                        | 1.15      |mg/l           |
|Cl-                     |                        | 275       |mg/l           |
|SiO2(aq)                |                        | 8.4       |mg/l           |
|Ca++                    |                        | 550       |mg/l           |
|Fe++                    |                        | 0.69      |mg/l           |
|Mg++                    |                        | 150       |mg/l           |
|Mn++                    |                        | 0.07      |mg/l           |
|SO4--                   |                        | 2160      |mg/l           |
|F-                      |                        | 0.3       |mg/l           |
|HPO4--                  |                        | 0.01      |mg/l           |
|HCO3-                   |                        | 878       |mg/l           |
|UO2++                   |                        | .16       |mg/l           |
|-------------------------------------------------------------------------------|
|Input Solid Solutions                                                          |
|-------------------------------------------------------------------------------|
| none                   |            |            |             |               |
|-------------------------------------------------------------------------------|
|SUPPRESSED SPECIES    (suppress,replace,augmentk,augmentg)    value            |
|-------------------------------------------------------------------------------|
| none                   |            |            |             |               |
|-------------------------------------------------------------------------------|
|OPTIONS                                                                        |
|-------------------------------------------------------------------------------|
| - SOLID SOLUTIONS -                                                           |
|   * ignore solid solutions                                                    |
|     process hypothetical solid solutions                                      |
|     process input and hypothetical solid solutions                            |
| - LOADING OF SPECIES INTO MEMORY -                                            |
|   * does nothing                                                              |
|     lists species loaded into memory                                          |
| - ECHO DATABASE INFORMATION -                                                 |
|   * does nothing                                                              |
|     lists all reactions                                                       |
|     lists reactions and log K values                                          |
|     lists reactions, log K values and polynomial coef.                        |
| - LIST OF AQUEOUS SPECIES (ordering) -                                        |
|   * in order of decreasing concentration                                      |
|     in same order as input file                                               |
| - LIST OF AQUEOUS SPECIES (concentration limit) -                             |
|     all species                                                               |
|     only species > 10**-20 molal                                              |
|   * only species > 10**-12 molal                                              |
|     not printed                                                               |
| - LIST OF AQUEOUS SPECIES (by element) -                                      |
|   * print major species                                                       |
|     print all species                                                         |
|     don't print                                                               |
| - MINERAL SATURATION STATES -                                                 |
|   * print if affinity > -10 kcals                                             |
|     print all                                                                 |
|     don't print                                                               |
| - pH SCALE CONVENTION -                                                       |
|   * modified NBS                                                              |
|     internal                                                                  |
|     rational                                                                  |
| - ACTIVITY COEFFICIENT OPTIONS -                                              |
|   * use B-dot equation                                                        |
|     Davies' equation                                                          |
|     Pitzer's equations                                                        |
| - AUTO BASIS SWITCHING -                                                      |
|   * off                                                                       |
|     on                                                                        |
| - PITZER DATABASE INFORMATION -                                               |
|   * print only warnings                                                       |
|     print species in model and number of Pitzer coefficients                  |
|     print species in model and names of Pitzer coefficients                   |
| - PICKUP FILE -                                                               |
|     write pickup file                                                         |
|   * don't write pickup file                                                   |
```

```
| - LIST MEAN IONIC PROPERTIES -                                      |
|   * don't print                                                     |
|     print                                                           |
| - LIST AQUEOUS SPECIES, ION SIZES, AND HYDRATION NUMBERS -          |
|     print                                                           |
|   * don't print                                                     |
| - CONVERGENCE CRITERIA -                                            |
|   * test both residual functions and correction terms              |
|     test only residual functions                                    |
|--------------------------------------------------------------------|
|DEBUGGING SWITCHES (0 = off, 1,2 = on                                |
|--------------------------------------------------------------------|
|0  generic debugging information                                     |
|0  print details of pre-Newton-Raphson iteration                    |
|0  print details of Newton-Raphson iteration                        |
|0  print details of stoichiometric factors                          |
|0  print details of stoichiometric factors calculation              |
|0  write reactions on RLIST                                          |
|0  list stoichiometric concentrations of master species             |
|0  request iteration variables to be killed                         |
|--------------------------------------------------------------------|
|DEVELOPMENT OPTIONS  (used for code development)                     |
|--------------------------------------------------------------------|
| none                                                                |
|--------------------------------------------------------------------|
|TOLERANCES               (desired values)      (defaults)           |
|--------------------------------------------------------------------|
|     residual functions  |                 |1.0e-10  tolbt          |
|     correction terms    |                 |1.0e-10  toldl          |
|     saturation state    |                 |0.5      tolsat         |
|number of N-R iterations |                 |30       itermx         |
|--------------------------------------------------------------------|
```

Data is delimited by the "|" symbol, and options are chosen by placing the "*" symbol opposite the appropriate description.

Note particularly the method for basis swapping. If for example we had wanted to have MW–12 saturated with quartz, rather than it having a silica concentration of 8.4 mg L^{-1}, and we wanted to control the CO_2 fugacity to 10^{-2} bar, we would swap quartz for the basis species $SiO_2(aq)$, and $CO_2(g)$ for the basis species HCO_3^-. In EQ3/6 this is done by entries in the BASIS SWITCH/CONSTRAINT column, as follows. It is a vital part of geochemical modeling to know how to perform this basis swapping, and to understand how and when to do it.

Electrical Balancing on	SO4--	code selects	not performed
SPECIES	BASIS SWITCH/CONSTRAINT	CONC/ETC	UNITS OR TYPE
redox		0.0	logfO2
H+		6.7	ph
Na+		265	mg/l
K+		14	mg/l
Al+++		1.15	mg/l
Cl-		275	mg/l
SiO2(aq)	Quartz	0.0	mineral
Ca++		550	mg/l
Fe++		0.69	mg/l
Mg++		150	mg/l
Mn++		0.07	mg/l
SO4--		2160	mg/l
F-		0.3	mg/l
HPO4--		0.01	mg/l
HCO3-	CO2(g)	-2	log fugacity
UO2++		.16	mg/l

Script Input

Script input is the most flexible format and perhaps the easiest to use. A file is prepared from scratch, or by altering a previous input file, and the formatting is relatively
unimportant. Extra spaces or empty lines, which can cause other programs to crash,
have no effect. Keywords may occur anywhere in the file, and may be upper or lower
case, although for readability it is common to use capitals for keywords, and to separate
keywords by empty lines.

In PHREEQC, a keyword signifies the beginning of a "data block", which may contain
several lines of data, and within which the ordering is important. After looking at a few
examples, it is not hard to make such a file. As in most programs, there is a facility in
the input file to change the data in the database, or to add new basis species, without
actually editing the database itself.

The MW–12 example used above would look like this:

```
TITLE Bear Creek MW12
SOLUTION 1
        units    mg/l
        pH       6.7
        temp     12.0
        redox    O(0)/O(-2)
        Ca                 550.0
        Mg                 150.0
        Na                 265.0
        K                   14.0
        Fe                   0.69
        Mn                   0.07
        Si                   8.4
        Cl                 275.0
        Al                   1.15
        S                 2160.0
        F                    0.3
        P                    0.01
        Alkalinity         878 as HCO3
        O(0)                 1.0      O2(g)    -0.7
        U(6)                 0.16
SOLUTION_MASTER_SPECIES
        U        U+4       0.0        238.0290      238.0290
        U(4)     U+4       0.0        238.0290
        U(5)     UO2+      0.0        238.0290
        U(6)     UO2+2     0.0        238.0290
SOLUTION_SPECIES
        #primary master species for U
        #secondary master species for U+4
        U+4 = U+4
                log_k         0.0
        U+4 + 4 H2O = U(OH)4 + 4 H+
                log_k        -8.538
                delta_h      24.760 kcal
        U+4 + 5 H2O = U(OH)5- + 5 H+
                log_k       -13.147
                delta_h      27.580 kcal
PHASES
        Uraninite
        UO2 + 4 H+ = U+4 + 2 H2O
        log_k        -3.490
        delta_h     -18.630 kcal
END
```

This example looks a little complicated, because the PHREEQC database does not
contain any data for uranium.[3] Therefore, it must be supplied by the user, either by

---

[3]This refers to the original PHREEQC database. A newer one, llnl.dat includes uranium species, so the
SOLUTION_MASTER_SPECIES and SOLUTION_SPECIES sections are not necessary.

editing the database, or, as here, by adding the data to the input file. Both the possible aqueous species, and their homogeneous and heterogeneous reactions should be included, with both $\log K$ and $\Delta_r H°$ (delta_h) data for each reaction.

Other examples of PHREEQC input files will be discussed in later chapters.

4.5.2 Program Output

There are fewer problems involved in understanding the output of geochemical modeling programs than in getting the input right. There may be some confusing features, such as reporting the speciation results twice – once for 25°C and once for the actual temperature, or once reporting all supersaturated phases and once reporting the result if all supersaturated phases precipitated out. However, a careful reading of the documentation is sufficient to understand the format. More tricky is the distinction between basis species and secondary species, which was discussed in Chapter 3, p. 49.

The results from speciation models are relatively simple. A common problem in reaction path models is their voluminous output, often amounting to tens of pages, depending on how many time-steps have been taken and what the print interval is. With so much data, it can be a problem to find exactly what is necessary, especially to get a graphical representation of just those variables which the user is interested.

In this respect, The Geochemist's Workbench™ is outstanding because a flexible graphical program is interfaced with the other programs in the package. A simple but effective alternative is provided by PHREEQC. In the input file, one of the keywords is used to tell the program which output data the user is especially interested in, and that is required in graphical form. The program then prints a second output file, with those data arranged in a tabular form suitable for importing into a spreadsheet or graphics program, in which the desired graphs can be prepared.

For example, here is a PHREEQC input file which would give results very much like those of Figure 8.6 (which was actually created with The Geochemist's Workbench™).

```
TITLE Recreate buffer diagram made with GWB

SOLUTION 1

temp 25
unit mol/kgw

pH 7.0
Na 1.0
Cl 1.0 charge
S(6) 1e-7
Si  1e-7

EQUILIBRIUM_PHASES

CO2(g)    -3.5
Fe(OH)3(a) 0.0 0.005 mol
calcite    0.0 0.08 mol
gibbsite   0.0 0.025 mol
gypsum     0.0 0.0

REACTION

H2SO4 1.0
.2 moles in 20 steps

SELECTED_OUTPUT
```

```
-file buffer1.pun
-a H+
-si calcite gibbsite Fe(OH)3(a) gypsum
-equilibrium_phases calcite gibbsite Fe(OH)3(a) kaolinite gypsum
```

The point here is the keyword SELECTED_OUTPUT. The data following this keyword cause a file named "buffer1.pun" to be created, containing, for each of the 20 steps performed, the pH (-a H+), the Saturation Indices for calcite, gibbsite, amorphous $Fe(OH)_3$, kaolinite, and gypsum (-si calcite ...), and the amounts of these phases created or destroyed (-equilibrium_phases calcite ...).

Part of this file looks like this (some columns have been omitted here):

```
step_x   ph      calcite   cc_d     gibbsite gibbsite_d  cc_si  gibbsite_si Fe(OH)3(a)_si

0.010    7.895    0.070    -0.010    0.025    0.000      0.000      0.000       0.000
0.020    7.756    0.060    -0.020    0.025    0.000      0.000      0.000       0.000
0.030    7.675    0.050    -0.030    0.025    0.000      0.000      0.000       0.000
0.040    7.618    0.040    -0.040    0.025    0.000      0.000      0.000       0.000
0.050    7.618    0.030    -0.050    0.025    0.000      0.000      0.000       0.000
0.060    7.618    0.020    -0.060    0.025    0.000      0.000      0.000       0.000
0.070    7.618    0.010    -0.070    0.025    0.000      0.000      0.000       0.000
0.080    5.671    0.000    -0.080    0.025    0.000     -3.894      0.000       0.000
0.090    3.940    0.000    -0.080    0.018   -0.007     -7.395      0.000       0.000
0.100    3.846    0.000    -0.080    0.012   -0.013     -7.619      0.000       0.000
0.110    3.793    0.000    -0.080    0.005   -0.020     -7.757      0.000       0.000
0.120    3.367    0.000    -0.080    0.000   -0.025     -8.641     -1.196       0.000
0.130    2.280    0.000    -0.080    0.000   -0.025    -10.844     -4.476      -2.670
0.140    1.811    0.000    -0.080    0.000   -0.025    -11.800     -5.895      -4.086
0.150    1.595    0.000    -0.080    0.000   -0.025    -12.251     -6.558      -4.748
0.160    1.453    0.000    -0.080    0.000   -0.025    -12.549     -6.994      -5.186
0.170    1.348    0.000    -0.080    0.000   -0.025    -12.773     -7.319      -5.513
0.180    1.265    0.000    -0.080    0.000   -0.025    -12.952     -7.578      -5.774
0.190    1.196    0.000    -0.080    0.000   -0.025    -13.102     -7.794      -5.992
0.200    1.137    0.000    -0.080    0.000   -0.025    -13.231     -7.979      -6.179
```

In this file, step_x is obviously the number of moles of H_2SO_4 added; calcite, gibbsite, etc. is the number of moles of each mineral remaining in the system (we abbreviated the word "calcite" to "cc" to save space); calcite_d, gibbsite_d, etc. is the number of moles of each mineral dissolved or destroyed; and calcite_si, gibbsite_si, etc. is the Saturation Index (see Table 3.2) of each mineral. These numbers are scattered throughout many pages in the normal output file, and are assembled here for convenient import into a spreadsheet or other program.

5

Preparation and Construction of a Geochemical Model

5.1 Introduction

To set up a geochemical model, we need:

- specific information describing the geological system of interest;

- conceptualization of what chemical reactions are occurring and what chemical reactions are important to the questions we seek to answer;

- thermodynamic, kinetic, and surface properties for the specific chemical system.

5.2 Establish the Goals

The goals of geochemical modeling will determine what type of models to develop and how detailed they need to be. They also determine what samples to collect and what parameters to measure. The purposes can range from establishing the baseline geochemistry or background concentrations, predicting contaminant fate and transport, and evaluating remedial alternatives. Usually, no matter what the ultimate goals are, there is a need to use geochemical modeling to characterize the dominant water–rock interactions at a site.

5.3 Learn the Groundwater Flow System

Some basic knowledge of the directions and rates of groundwater flow at a site is important for deciding the sample collection and model conceptualization. The direction of groundwater flow determines the sequence in which the water will contact different mineral assemblages in the aquifers. Knowledge of the flow path ensures that observed chemical variation results from a evolutionary path, and this variation can be used in our conceptualization of chemical reactions in an aquifer. For example, knowledge of the flow paths is essential for the application of inverse mass balance modeling (see Chapter 9). The rate of groundwater flow determines, for example, whether or not the local equilibrium assumption can be applied.

However, our understanding of the groundwater flow patterns is often an iterative process. Chemicals and isotopes can be excellent indicators of the directions and rates of groundwater flow. One goal of geochemical modeling can be to help determine the directions and rates of groundwater flow.

5.4   Collection of Field and Laboratory Data

In order to model chemical reactions in a specific system, we need to know the *geochemical properties of the aquifer*. Any study of fluid/sediment/mineral interactions requires knowledge of both chemical compositions of fluids and mineralogical compositions of the solid matrix.

5.4.1   Decide Which Parameters to Measure for Groundwater

In many groundwater studies, only constituents of environmental concern, e.g. metals that are regulated, are analyzed in the laboratory (Davis, 1988). However, if the purpose of collecting and analyzing groundwater samples is to perform geochemical modeling, constituents and parameters that are not of environmental concern, but that are essential for characterizing water–rock interactions in the aquifer, also need to be analyzed. That typically would include pH, water temperature, dissolved oxygen, alkalinity measurements in the field, and laboratory analyses of Na, K, Ca, Mg, Al, Fe, Mn, SiO_2, Cl, SO_4^{2-}, and NO_3^- (Table 5.1).

The list should also include constituents that are not typically in high concentrations (e.g., > 1 mg L^{-1}) in natural environments, but possibly occur in high concentrations because of anthropogenic inputs at a specific site. Prudent geochemists usually ask the laboratories to "scan" all the constituents in a couple of samples or all samples for the first round of sampling to detect any ions with significant concentrations.

5.4.2   Characterize the Solids

In order to perform geochemical modeling and interpret the results, the composition and properties of the solid media also should be characterized (Penn *et al.*, 2001; Zhu and Burden, 2001). This unfortunately is often not included in a "groundwater study". Much could be said about the millions of dollars expended on groundwater analyses without a single analysis of the minerals at a site.

The mineral compositions and the abundances of the primary and secondary minerals in a setting are important for interpreting the solubility control on groundwater chemistry. This information can be obtained by conducting X-ray diffraction (Cullity, 1978) and petrographic studies (Ineson, 1989; Kerr, 1977), scanning electron microscopy (SEM; see Goldstein *et al.*, 1992; Reed, 1996), electron microprobe (Goldstein *et al.*,

	Field measurements
Water	temperature
	pH
	alkalinity
	dissolved oxygen

	Laboratory measurements
Water	Na, K, Mg, Ca, Al, Fe, Mn
	Cl, HCO_3^-, SO_4^{2-}, NO_3^-, SiO_2
Solids	mineral identification
	mineral abundance
	cation-exchange capacity (CEC)
	extractable Mn and Fe oxides
	soil pH
	acid-generation potential
	acid-neutralization capacity

Table 5.1. Typical parameters measured for geochemical modeling.

1992; Reed, 1996), and, for fine-grained materials, transmission electron microscopy (TEM) or high resolution TEM (HRTEM; see Spence, 1988; Veblen, 1990; Williams and Carter, 1996).

Cation-exchange capacity (CEC) is another parameter that should be measured (Jackson, 1985; Sumner and Miller, 1996) if ion-exchange between groundwater and clay minerals is suspected to exert a control on the groundwater geochemistry and contaminant migration. It is essential to know the amounts of amorphous iron and manganese oxides, because of their high adsorptive capacity, if surface complexation modeling is to be performed. This information can be obtained by conducting sequential extraction analyses using the method by Chao (1972), Chao and Zhou (1983), and Loeppert and Inskeep (1996).

If acid migration is the primary concern, the acid-neutralization capacity of the aquifer matrix becomes an important parameter. The analytical procedures used to measure this potential are well known (Sobek *et al.*, 1978). Similarly, the acid-generation capacity of a rock is also an important parameter. Acid-generation potential is usually defined as the potential release of acid when the sulfides in a rock matrix are completely oxidized (Sobek *et al.*, 1978). However, in a completely oxidized rock or soil matrix, such as some uranium mill tailings, a significant amount of acid can exist on the mineral surfaces and on exchangeable sites when they have been in contact with acidic water (Rose and Gahzi, 1997; Rose and Elliott, 2000; Zhu *et al.*, 2002). This acid can be released and can generate a long term source of acid.

5.4.3   Evaluate Quality of Water Analyses. Charge Balance I

The quality of field and laboratory measurements varies. A common criterion for determining the quality of water analyses is the charge balance among the reported cation and anion concentrations (Freeze and Cherry, 1979). The principle of electroneutrality

Element	Species	Conversion factors		Element	Species	Conversion factors	
		A	B			A	B
Aluminum	Al^{3+}	0.11119	0.03715	Lead	Pb^{2+}	0.00965	0.00483
Ammonium	NH_4^+	0.05544	0.05544	Lithium	Li^+	0.14407	0.14407
Barium	Ba^{2+}	0.01456	0.00728	Magnesium	Mg^{2+}	0.08229	0.04114
Beryllium	Be^{2+}	0.22192	0.11096	Manganese	Mn^{2+}	0.03640	0.01820
Bicarbonate	HCO_3^-	0.01639	0.01639	Molybdenum	Mo	—	0.01042
Boron	B	—	0.09250	Nickel	Ni^{2+}	—	0.01704
Bromine	Br^-	0.01252	0.01252	Nitrate	NO_3^-	0.01613	0.01613
Cadmium	Cd^{2+}	0.01779	0.00890	Nitrite	NO_2^-	0.02174	0.02174
Calcium	Ca^{2+}	0.04990	0.02495	Phosphate	HPO_4^{3-}	0.03159	0.01053
Carbonate	CO_3^{2-}	0.03333	0.01666	Potassium	K^+	0.03159	0.01053
Cesium	Cs^+	0.00752	0.00752	Rubidium	Rb^+	0.01170	0.01170
Chlorine	Cl^-	0.02821	0.02821	Silica	$SiO_2(aq)$	—	0.01664
Cobalt	Co^{2+}	0.03394	0.01697	Silver	Ag^+	0.00927	0.00927
Copper	Cu^{2+}	0.03147	0.01574				
Fluorine	F^-	0.05264	0.05264	Sodium	Na^+	0.04350	0.04350
Hydrogen	H^+	0.99216	0.99216	Strontium	Sr^{2+}	0.02283	0.01141
Hydroxyl	OH^-	0.05880	0.05880	Sulfate	SO_4^{2-}	0.02082	0.01041
Iodine	I^-	0.00788	0.00788	Sulfide	S^{2-}	0.06238	0.03119
Iron	Fe^{2+}	0.03581	0.01791	Zinc	Zn^{2+}	0.03059	0.01530
Iron	Fe^{3+}	0.05372	0.01791				

Table 5.2. Conversion factors from $(mg\ L^{-1}) \times A$ to $(meq\ L^{-1})$ or vice versa with B. Reproduced from Hem (1985, Table 9).

requires that the ionic species in an electrolyte solution such as groundwater maintain a charge balance on a macroscopic scale. A large charge imbalance between reported cations and anions indicates either that a major anion or cation has not been included in the analysis, or that there were problems associated with sampling, sample preservation and handling, or laboratory errors.

Charge imbalances are commonly reported as

$$\text{imbalance}\% = \frac{\sum \text{cations} - \sum \text{anions}}{\left(\left| \sum \text{cations} \right| + \left| \sum \text{anions} \right| \right)} \times 100 \qquad (5.1)$$

The concentrations used in Equation (5.1) are in milliequivalents per liter (meq L^{-1}), which can be calculated from milligrams per liter (mg L^{-1}) using conversion factors listed in Hem (1985). See Table 5.2.

The cut-off value for acceptable charge imbalance is empirical and somewhat arbitrary. Freeze and Cherry (1979) reported that analytical laboratories usually consider a charge-balance error of $< 5\%$ to be acceptable. However, for dilute solutions such as rain water, the errors are usually higher.

Fritz (1994) surveyed the water analyses published in 68 articles in *Applied Geochemistry*, *Geochimica et Cosmochimica Acta*, *Water Resources Research*, *Ground Water*, and *Journal of Hydrology*. Many reported water analyses contain large charge-balance errors. From our experience, the percentages of water analyses having unacceptable charge-balance errors are even larger in unpublished environmental reports. For example, Zhu and Wille (1997) examined all water analyses at Frenchman Flat, the Nevada Test Site, and found that among the 160 "complete" major ion analyses,

---

Stoichiometric Charge Imbalance

Let's look at an acidic water sample ($pH = 3.8$ and major ions listed below). PHREEQC calculations show a perfect charge balance in the speciated solution. However, calculations using a spreadsheet show about 5 % stoichiometric charge imbalance.

Element	(mol L^{-1})	Charge	(meq L^{-1})
Al	3.86×10^{-2}	3	115.9
Cl	1.59×10^{-2}	-1	-15.9
Fe(3)	3.57×10^{-2}	3	107
Mg	4.20×10^{-2}	2	84.1
Na	9.63×10^{-2}	1	96.3
S(6)	1.76×10^{-1}	-2	-351
H	1.58×10^{-4}	1	0.2
	sum =		36.5
	out of balance =		5 %

Thus, speciated and stoichiometric charge balances can be quite different, except in dilute solutions. Note, too, that the concentrations of H^+ and OH^- (as determined from the pH) should be routinely included in the stoichiometric charge balance, even though they will only be significant in very acidic or very basic solutions.

---

100 analyses (63%) have charge imbalances exceeding $\pm 5\%$. Analyses showing charge imbalance exceeding 50% were common. Most of the problems appeared to come from field alkalinity measurements.

Note that the charge imbalances calculated from Equation (5.1) are "stoichiometric" charge imbalances, which are based on the assumption that all measured elemental concentrations appear as charged ionic species, for example that all aluminum in the solution is Al^{3+}; all the calcium is Ca^{2+}; and so on. This stoichiometric charge-balance calculation tends to work well as a screen for checking the quality of dilute groundwater and surface water analyses, which are largely uncomplexed (see Hem, 1985).

Analytical results that show a high charge imbalance can be closely charge balanced when aqueous speciation is considered. At many contaminated sites, such as acid mine drainage sites, aqueous complexes that bind with H^+ and OH^- are important. Some proportion of the ions occurs in the solution as neutral species and complexed ions. In the Box above, we give an example where hydroxyl complexes are important and a perfectly charge-balanced solution, according to the speciation model, can have a large stoichiometric charge-imbalance. MINTEQA2 routinely performs a charge imbalance check for both the unspeciated and the speciated analysis.

But the charge-balance situation is even more complicated. To see this, consider the following points.

- Water analyses report the pH and total concentrations of the ions as free ions, even though they may exist in complexed forms. For most complexes, this does not change the calculated charge balance. For every complex that forms from a cation, it reduces the cation charge but also reduces the anion charge. For example, the formation of $AlSO_4^+$ reduces an equal meq of Al^{3+} and SO_4^{2-} from the charge pool. The charge balances in stoichiometric and speciated solutions remain unchanged.

- For the formation of hydroxyl and proton complexes, the story is different. Aluminum, for example, may exist in solution dominantly as various hydroxy-complexes such as $Al(OH)_2^+$ and $Al(OH)^{2+}$. Obviously, then, the charge on much of the aluminum in solution is less than +3. But the OH^- contribution to the charge balance is determined from the pH, which reflects only the *free* OH^-, not the complexed OH^-. In acid solutions, free OH^- are completely insignificant, but the complexed OH^- is not. In very alkaline solutions, the story is similar, but it is the role of H^+ which becomes important. It is this factor which causes the difference between stoichiometric and speciated charge balances.

- The concentrations of hydroxyl and proton complexes which affect the charge balance are model-dependent. Basically, this means that just as the stoichiometric charge balance depends on the quality of the analysis, the speciated charge balance depends on the quality of the thermodynamic database. Programs using different databases will produce different charge balances, and sometimes the difference can be large.

- Therefore, it may be that a water analysis which is error free can still have a large speciated charge imbalance if the database is of poor quality, i.e., if the equilibrium constants are not accurate or not internally consistent.

Of course, charge balance is not the only criterion for evaluating the accuracy or quality of chemical analyses of water samples, but it is a good first screening. Large charge-balance errors always tell us that something is wrong. However, a good charge balance does not necessarily mean that all is well – Freeze and Cherry (1979) point out situations where large errors in the individual ion analyses balance one another. Comparison with historical data also serves as a good method to spot errors. The modeler also should pay attention to data that seem to have unusual concentrations given the site geology.

5.5   Decide What Types of Model to Construct

Before constructing a detailed chemical model, we need to make the following decisions:

1. what components are to be included in the model;

2. what types of geochemical reactions are to be included in the model;

Stoichiometric versus Speciated Charge Imbalance: Example 2

Now, let's look at a real analysis reported by White *et al.* (1984). The tailings pore fluid sampled at the Riverton uranium mill tailings site in Wyoming is very acidic and contains high total dissolved solids. The table below shows the analysis.

Element	(mol L^{-1})	Charge	(meq L^{-1})
Al	0.056	3	167.9
Cl	0.00601	-1	-6.0
Fe(3)	0.0307	3	92.1
Mg	0.0457	2	91.3
Na	0.0117	1	11.7
S(6)	0.183	-2	-366.4
K	0.00000691	1	0.00691
H (*pH*= 2.3)	0.00501	1	5.0
	sum =		10.6
	out of balance=		1.4 %

Calculation shows a 1.4 % stoichiometric charge imbalance. However, PHREEQC calculations using the default database phreeqc.dat result in a 7.29 % charge imbalance in speciated solution. Using the database llnl.dat gives a charge imbalance of -0.06 %.

The differences between the stoichiometric and speciated charge balance are due to the formation of OH^- complexes as in the preceding example. The difference between the two speciated solutions result from the different species included and equilibrium constants used in the databases. The two databases give vastly different species distribution for Fe^{3+} (Figure 5.1).

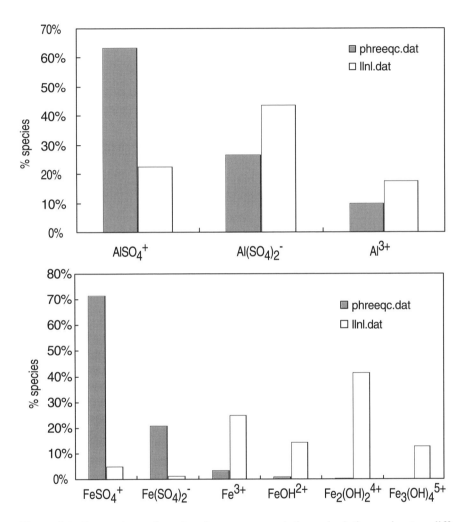

Figure 5.1. Comparison of results of PHREEQC speciation calculations using two different databases for the tailings fluid from the Riverton uranium mill tailings site (White
et al., 1984).

3. whether to include kinetics or assume local equilibrium; and

4. what other processes are to be simulated.

Some conceptualization of chemical reactions at the site is necessary for successful modeling. The development of a conceptual model may occur after careful examination of field data or preliminary modeling results. It is often difficult to identify which reaction is responsible for the observed chemical concentration distributions. A combination of general knowledge of geochemistry and the site geochemistry is needed to make a decision.

It is unnecessary, and in fact impossible, to include all chemical components, all types of chemical reactions, or all chemical reactions in the aquifer system in the model. The configuration of a chemical model, decisions on the chemical components, reaction types, and reactions to be included, should depend on site-specific conditions and what questions we want to ask.

For a given geological system, i.e., a particular site, different concerns define different geochemical models. For example, at an abandoned uranium mill tailings site, we may have concerns that both sulfate and uranium exceed the regulated limits at the point of compliance. However, geochemically, we need to consider chemical reactions in two different situations for sulfate and uranium. Sulfate is a major component in groundwater. Its concentration in groundwater is likely to be controlled by solubility of common mineral phases like gypsum or jarosite. So we only need to consider major chemical components and pH in the groundwater, together with dissolution–precipitation reactions. There are numerous other chemical reactions that occur at the site, but we can safely ignore them because they do not significantly affect the fate and transport of sulfate.

However, when considering uranium transport at the same site, we need to consider a different set of chemical reactions. Uranium is a trace component in groundwater. The consideration of the interactions between the chemical components, both in the aqueous phases and between the aqueous and solid phases, is different from sulfate, the major component. Uranium transport is most likely controlled by surface adsorption onto mineral surfaces. So the surface adsorption reactions that are safely ignored in the sulfate case are now important. Surface reactions concerning other trace metals, for example, Ni and Cd, are ignored in the sulfate transport case, but now they need to be included because of competitive adsorption. Furthermore, even though our concern is about a trace component, i.e., uranium, we need to know the major components of groundwater to study uranium transport because the iron oxide surface sites may be mostly occupied by sulfate at low pH.

Our knowledge of some geochemical systems is decidedly better than others. Therefore, the lack of reaction properties may be a prohibitive problem for one chemical system, but not for another where the contaminant of concern is different. Thus, what types and what models are used depends on the goals of modeling.

5.6   Gather Chemical Properties

Next, the modeler needs to assemble the thermodynamic, kinetic, and surface properties that describe chemical reactions for modeling. These properties are not site-specific, but chemical system specific.

For aqueous speciation modeling, the equilibrium constants for aqueous complexes are needed. The modeler particularly needs to check whether the elements of interest are in the database at all, and, if not, how to add the master species or basis species. (See the Appendix for a detailed discussion of how to add a basis species and secondary species into a database.) Sometimes the reaction enthalpies are needed for corrections of the temperature effects if reactions are modeled at temperatures different from 25 °C.

For solubility calculations, the thermodynamic properties for the relevant minerals are needed. These properties are often given as equilibrium constants of the formation or dissolution reactions.

For modeling surface adsorption using the surface complexation theory, we need properties of the surfaces as well as complexation constants for the sorbant. Surface properties include site density, surface areas, and molecular formula weight. If we use the triple layer model, capacitance data are also needed; see Chapter 7 for more details.

Models that include kinetics also need kinetic data. Properties for kinetic modeling include the rate constants, effective surface areas (e.g., Murphy and Helgeson, 1988), and the forms of rate laws.

The first stop for the modeler is to take a good look at the database that accompanies the code or codes. The modeler needs to examine what thermodynamic and kinetic parameters are included in the database. The database may or may not contain the thermodynamic and kinetic properties for the chemical species we want to include in the model, and the listed properties may not be internally consistent. See Chapter 4 for more details on the thermodynamic data and databases.

5.7   Select a Computer Code

In choosing a computer code, one should consider the following.

- The type of processes the code is capable of simulating. For example, the capability of simulating surface complexation and ion-exchange will dictate the choice of code. Table 4.2 lists the type of chemical reactions that the most popular codes can simulate.

- The availability and accuracy of the thermodynamic data in the code's database. This is seemingly an obvious criterion, but it is difficult to apply. Database evaluation is a specialized task that is beyond the ability or the interest of most users. This subject is discussed further later in this chapter.

- The ease of use. Codes for which the input file is easy to prepare and the output files are easy to understand are of course preferred. Otherwise, the codes are difficult to use and mistakes can be made.

5.8   Set Up a Model

5.8.1   Basis Swapping

Modeling programs assume that the user will assign concentrations or activities to some number of basis species. These selected basis species then become a part of the basis, the list of species used to define the composition of the system. As mentioned earlier (p. 52), if a mineral or a gas is part of the system, it controls the activity and hence the concentration of one basis species, and must be included in the basis in place of that species. The process of doing this is called "swapping" or "switching" the mineral or gas into the basis. All modeling programs provide some way of doing this, and the method used by EQ3/6 was demonstrated in §4.5.

Secondary species can also be swapped into the basis in place of basis species. This might be desirable if the original basis species becomes extremely small in concentration, causing numerical difficulties. For example, if the aluminum basis species is Al^{3+}, it would be desirable to swap $Al(OH)_4^-$ for Al^{3+} before beginning the calculation for most natural waters, because it probably exists in concentrations orders of magnitude greater than Al^{3+}, and convergence will be helped.

During the calculation, minerals may dissolve completely or precipitate, and therefore need to be swapped in or out of the basis by the program as the calculation proceeds. Bethke (1996) shows clearly how this is accomplished using the techniques of linear algebra.

5.8.2   Charge Balance II

Because all real solutions of electrolytes have a balance of positive and negative charges, a perfect chemical analysis will result in calculated ionic species which are also charge balanced. But chemical analyses invariably contain errors of precision and accuracy, and sometimes an important element is omitted from the analysis, resulting in calculated ionic species which are not charge balanced. Practical aspects of the charge balance are considered in § 5.4.3. Here, we want to point out that most modeling programs provide the option of forcing a balance of charges by adjusting the concentration of a chosen basis species. As Bethke (1996) points out, Cl^- is often chosen, because it is usually abundant, and its concentration may not be well known. In addition, there are very few chloride minerals in most databases, so changing m_{Cl^-} will not greatly change any Saturation Indices (it will have some small effect through the effect of ionic strength on the activity coefficients).

The point here is that the charge-balance option is a replacement for the choice of a basis species concentration. Thus, in our discussion of the Phase Rule, each time we mentioned b basis species, we could have said $b-1$ basis species *plus* the charge-balance condition on an additional basis species.

Note, however, that it is not a good idea to balance on H^+ in order to calculate the pH, unless it is a hypothetical model solution with perfectly known concentrations. This is because the concentration of H^+ is such a small quantity that tiny errors in the analytical concentrations of the other elements will render the calculated pH meaningless. Choosing Cl^- as the balancing species is usually fine, unless the solution has a negative imbalance greater than the reported amount of Cl^-. In this case, some cation must be chosen, typically Na^+, and to a lesser extent Ca^{2+}. However, changing the concentration of some species will affect many activity coefficients and Saturation Indices, and we don't want to change the concentration of a species that we are fairly sure is accurately known. If the imbalance is so great that problems are created no matter what choice is made, it might be better to reanalyze the solution.

Finally, we might note that, although forcing a charge balance is perhaps customary, and may aid convergence in some cases, there may be cases where it should not be used. For example, we may have an analysis which is perfectly good for those elements which were analyzed, but which missed some important organic acids. Changing the concentration of some perfectly well known species in this case does not help the situation. In other cases, it might be wise to balance on different elements, to see which seem most plausible in light of the overall problem.

5.9   Interpretation of Modeling Results

We will discuss the modeling interpretations in the following chapters. But we should always ask ourselves the question "Is the model sufficient or accurate?" Bethke (1996) drafted a check list for geochemical modeling which we have modified somewhat and present here.

1. Is the chemical analysis used sufficiently accurate to support the modeling study?

2. Does the thermodynamic database in the program contain the aqueous species, surface complexes, and minerals likely to be important in the solution?

3. Are the equilibrium constants for the important reactions in the thermodynamic database sufficiently accurate to warrant the conclusions?

4. Is the ionic strength small enough so that the activity coefficient model in the modeling code is applicable?

We will elaborate on a few points related to the above questions.

5.9.1   Accuracy and Completeness of the Database

The fact that a code has an accompanying thermodynamic database makes the modeler's life much easier. Unfortunately, if a modeler has never used a code that does not have a database, he or she could get the impression that a modeling program is the same thing as a model. This confusion is enhanced by the titles of some programs such as "so-and-so program – A so-and-so model." That a model is produced by a modeling program is true to a certain extent, but there are some misconceptions about codes, models, and thermodynamic databases, which often mean different things.

For example, a species distribution and mineral solubility model generated by a program depends on the values of the thermodynamic properties used for the species and solids involved. The equilibrium constants for amorphous ferric iron hydrous oxide vary by several orders of magnitude (Stumm and Morgan, 1981). By using one value, the program produces a model that says a solution is supersaturated with respect to iron oxide; but with another value it may produce a model saying that the solution is undersaturated with iron oxide.

Sometimes, the availability or absence of equilibrium constants for certain aqueous species and solids in the database produces incorrect modeling results or models. For example, most databases do not have data for metal–organic complexes. Even for a solution with high organic concentrations the codes will show no metal–organic complexes because of this absence, although in fact these complexes might be significant in the real system. Similarly, for a solution with high Al and sulfate concentrations, the solubility controls on Al^{3+} and SO_4^{2-} may not be correctly modeled because no equilibrium constants are available for Al–sulfate phases in the database.

5.9.2 Input Constraints

Besides the accuracy and completeness of the thermodynamic data, a model is also produced with certain input constraints, e.g., analytical concentrations of groundwater samples. This means that the accuracy and completeness of analytical data also influence a model. For example, if the water is predominantly sulfate-rich and the analysis failed to analyze sulfate, then speciation and solubility models cannot be correct because a major component was missed in analysis. This shortcoming has nothing to do with the code or program itself.

Programs typically give the modeler various choices in the input file, e.g., choices relating to redox states. Different choices may result in different speciation and solubility models.

In the case of reaction path and inverse mass balance calculations, the involvement of the modeler in producing a geochemical model is crucial. Calculations may indicate that dozens or hundreds of minerals are supersaturated in a system. To perform reaction path calculations, the modeler has to use geochemical, geological, and mineralogical knowledge to decide which are the most likely phases to exist in the particular geologic environment of interest.

5.9.3 Who Produced the Model?

The common perception among environmental practitioners, that a modeling program produces a model, is not correct. *It is always the modeler, not the computer code, that produces a model.* In environmental practice, this misconception, and sometimes misrepresentation, borders on deception of the public, regulators, or clients because most popular modeling codes are distributed by regulatory or federal agencies and therefore carry an air of authority.

In summary, a model is invariably produced with certain input from the modeler, even though the need for conceptualization on the modeler's part is not always apparent. The modeling programs and databases have *not* been tested for the accuracy and completeness of their thermodynamic properties. Sometimes, running the code without worrying about the underlying conceptualization produces acceptable modeling results. Often, however, we are not that lucky, and it is the modeler's responsibility to ensure that the results are produced from an acceptable chemical model.

5.10   Reporting and Presentation of Modeling Results

Modeling work is not complete until it is effectively presented either orally or in a report. Often, managers or regulators, who review the geochemical modeling work, are not expert on the subject. Clear presentation is the key.

A presentation of modeling work needs a description of the conceptual model, the assumptions in the model, sources of thermodynamic and kinetic data, comparison of modeling results with filed and laboratory data, and model predictions. Some simple descriptions of the modeling code are also needed. Model limitations should be clearly stated.

A word of advice to modelers from our own experience is perhaps in order: *keep a modeling journal*. During the course of a modeling study, many changes in parameters or even codes are made. One soon loses track of all the details. Why did I use that value? Where did it come from? It can be months if not years after the modeling study is completed until one hears back from the regulatory agencies or editors of the journal to which one has submitted a manuscript. Date and label all figures and tables.

6

Speciation and Solubility Modeling

6.1 Introduction

The application of speciation–solubility in geochemistry goes back to Garrels and Thompson (1962), who calculated the aqueous speciation in seawater and saturation states with respect to mineral solubilities. Since then, this subject has been treated extensively, and hundreds of codes are available for this kind of calculation. In principle, the concentration and activity coefficients of all aqueous species can be calculated if there are the same number of equations as unknowns. Given an equilibrium constant for each complex, plus an analysis giving the total quantity of each basis species, this is usually possible.

Speciation–solubility calculations provide a "snapshot" of the (assumed) equilibrium state of a dynamic system. That is, the chemical composition of a water sample is obtained, and assuming that the chemical species in the solution are at mutual equilibrium (*homogeneous equilibrium*), the concentrations and activities of the various ionic and molecular species present are calculated. Then, using these calculated quantities, the saturation states (i.e., whether superstaurated or undersaturated) of all possible pure solids and gases are calculated, as described in Chapter 3. Here, "all possible" refers to the fact that the range of solids and gases that can be calculated to exist in the system is limited by the chemical components in the water analysis. One cannot infer anything about the saturation state of calcite (or any other mineral containing calcium) if the water has not been analyzed for calcium.

The interpretation of calculated mineral Saturation Indices is not simple, and is discussed in §6.2.7.

Figure 6.1.   Overview of the Bear Creek Uranium tailings site, looking toward the south.   The white buildings on the left, now demolished, indicate where the uranium mill operated from the 1970s to the mid-1980s.   Spent acids and tailings slurries were piped to tailings ponds behind the tailings dams.   This site has now been reclaimed.

6.2   A Uranium Mill Tailings Impoundment

6.2.1   The Site

The Bear Creek Uranium site is located in the southern part of the Powder River basin in Wyoming, and presents an example of uranium mill tailings and acid mine drainage (AMD) problems typically associated with mine tailings. A uranium mill operated from the 1970s to the mid-1980s. Sandstone uranium rollfront ores were processed at the mill using sulfuric acid and sodium chlorate to dissolve and oxidize uranium. Spent acids and tailings were piped in a slurry to unlined tailings ponds (Figure 6.1). The tailings fluid has a pH between 1.5 and 3.5, and a total dissolved solid (TDS) concentration close to $20\,000$ mg L^{-1}. A number of toxic metals and radionuclides are present in the tailings fluid at hazardous concentrations: arsenic (As), beryllium (Be), cadmium (Cd), chromium (Cr), lead (Pb), molybdenum (Mo), nickel (Ni), selenium (Se), radium (^{226}Ra, ^{228}Ra), thorium (^{230}Th), and uranium (^{238}U, ^{234}U). Seepage from the disposal ponds into the underlying shallow sandstone and alluvium aquifer has formed an acid plume in groundwater.

A seepage control system was installed which consisted of pumping water from wells into the tailings basin to control the further migration of the groundwater plume. The planned reclamation is to add a low-permeability cover on the tailings ponds to control

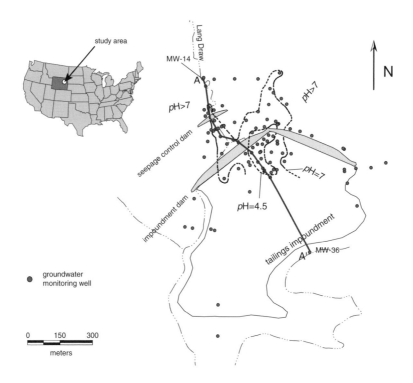

Figure 6.2. Plan view of the Bear Creek mine site and tailings impoundment.

recharge and retard further seepage into the aquifer. Hydrological modeling shows that it will take five years for the tailings fluid to drain after the low-permeability cover is installed. The aquifer contains calcite. The success of the reclamation plan depends on whether the aquifer can neutralize the acid contained in the low *p*H groundwater plume and attenuate the migration of the hazardous constituents. In other words, success depends on whether "natural attenuation" will sufficiently contain the contamination at the site once the source is terminated (see the Box on p. 110).

Figure 6.2 is a plan view of the site, showing the extent of the tailings pond, the locations of the tailings dam and seepage control dam, the size of the low *p*H groundwater plume, and the location of monitoring wells from which samples are available. Figure 6.3 is a cross-section along the line A–A′ in Figure 6.2, which is along a flow path toward the northwest. The two flow directions at the site are controlled by the site geology and delineated by the low *p*H groundwater.

6.2.2   The Purpose of Geochemical Modeling

The ultimate question for the parties and regulators responsible for the site (and the question that geochemists and hydrogeologists are hired to answer) is *what will be the concentrations of the regulated constituents at the point of compliance (POC) over the compliance period of 200 years?*

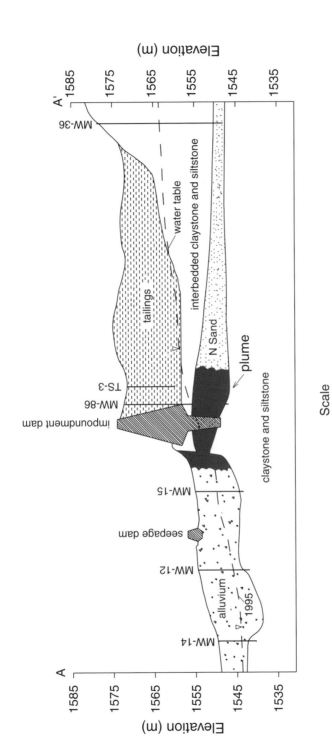

Figure 6.3. Cross-section A–A′ from Figure 6.2. Water level was recorded on January 5, 1995.

Uranium Mill Tailings

Uranium ore was mined in the USA in significant quantities by private companies to produce nuclear weaponry. After the 1950s, uranium was also needed as fuel for nuclear power plants to produce electricity. Ores are first crushed into centimeter to millimeter size. The ground ore is then disposed to the leach tanks to be dissolved and oxidized, respectively, by sulfuric acid (H_2SO_4) and sodium chlorate ($NaClO_4$). The dissolved uranium is extracted from the solutions.

The barren sand and slimes left after the leaching processes are called tailings. They contain about 85% of the radioactivity present in the unprocessed uranium ore. The tailings, spent acid, and process water are often pumped to a tailings impoundment. The tailings fluids or leachate from the tailings piles contain concentrations of radioactive and toxic materials that might pose a public health hazard.

The US Environmental Protection Agency (US EPA) developed standards to protect the public and the environment from potential radioactive and non-radioactive hazards at abandoned uranium ore processing sites (40 CFR 192). It is estimated that there are about 160 million cubic meters of mill tailings stored in temporary repositories in the USA (Miller *et al.*, 1988), and the cost for reclamation and remediation is estimated to be billions of dollars. In Canada, where a large percentage of the world's uranium was produced, it will cost about $3 billion to remediate groundwater contamination (Feasby *et al.*, 1991). Argentina, Australia, Germany, the Republic of South Africa, and Russia also have serious uranium mill tailings problems.

A number of regulations govern the concentrations that are allowed to be discharged outside of the property. These include the Nuclear Regulatory Commission's and US EPA's standards (Federal Registry 40 CFR 192), and State of Wyoming regulations for livestock and ecosystems. Recognizing the technical impracticality of controlling radionuclides at the regulated low concentrations and the fact that many uranium mill tailings impoundments are located in remote areas far from residential populations, the Nuclear Regulatory Commission allows responsible parties to petition for an Alternative Concentration Limit (ACL).

The concentrations of the hazardous constituents at the POC in the next 200 years will be the result of physical transport by the groundwater flow and chemical reactions between groundwater and the aquifer matrix. Geochemical reactions in the aquifer will obviously greatly influence the migration of the acid plume and the fate and transport of the hazardous constituents. To predict the contaminant transport accurately, these geochemical reactions should be incorporated into the transport model. The purpose of geochemical modeling, then, is to learn what geochemical reactions are possibly

---

Natural Attenuation

The reclamation and remediation of many contaminated sites, and uranium mill tailings sites in particular, have relied on "natural attenuation" processes. A number of physical, chemical, and biological processes can reduce the concentrations of contaminants.

- *Physical.* Dispersion, diffusion, and mixing with another water body will dilute the contaminant concentrations.

- *Chemical.* Precipitation, co-precipitation, ion-exchange, and surface adsorption can transfer contaminants from groundwater to the immobile solid phases and reduce the contaminant concentrations.

- *Biological.* Microbial mediated processes can transform organic contaminants into benign compounds or oxidize or reduce the toxic metals into a less toxic and mobile form.

A number of uranium mill tailings sites rely on "natural flushing", allowing natural groundwater movement and geochemical process to decrease contaminant concentrations. This was relied on at many of the US Department of Energy's UMTRA sites (http://www.doegjpo.com/gwwp/, October 25, 2000). This is also the strategy used at the Bear Creek site.

A scientific basis must be provided before natural attenuation can be accepted as an effective remediation alternative, because the public and communities often suspect this to be just a "do nothing" or "walk away" strategy by the polluters (National Research Council, 2000). The specific natural attenuation processes must be identified and quantified, and their long term sustainability must be evaluated (National Research Council, 2000). The geochemical modeling work, described in this section, provides the conceptual framework, which can help document and interpret the "footprints" (National Research Council, 2000) of the natural attenuation processes that are actually occurring in the aquifer.

---

occurring at the site, to evaluate the effects of geochemical reactions on contaminant transport, and ultimately to incorporate the results into the transport model that predicts the concentrations at the POC over the next 200 years.

6.2.3   Site Geology and Data

Numerous site monitoring data have been used to delineate four pH zones in the contaminated groundwater (Figure 6.2). The four zones have pH values of approximately 3.5, 4.5, 6.5–7.4, and ≥ 7.4, respectively (see discussion below). In designing the ground-

Analyte	Wells					
(mg L^{-1})	MW-14	MW-12	MW-15	MW-86	TS–3	MW-36
Al	< 0.5	1.15	1.33	230	1020	0.01
Ca	440	550	650	420	310	158
Fe	0.02	0.69	2.18	926	1950	0.01
Mg	59	150	250	700	1000	21
Mn	0.49	0.07	0.35	35.9	66.3	0.11
K	2	14	18	42	60	17
SiO$_2$	9.7	8.4	9.7	10	40.5	5.6
Na	34	265	212	278	89	61
Alkalinity						
as calcite	331	878	1450	<5	<5	153
SO$_4^{2-}$	1053	1500	1650	8100	16500	425
Cl	90	275	375	400	550	25
F	0.3	0.3	0.3	<0.1	<0.1	0.2
PO$_4^{3-}$	< 0.01	0.01	<0.01	0.01	0.01	<0.01
Temp. (°C)	13	12	13	12	16	15
pH	7.7	6.7	6.5	4.5	3.8	7.4

Table 6.1. Chemical analyses of samples from monitoring wells. The analytical data show a large excess of anions in the charge balance.

water sampling, we chose samples from locations upgradient of the acid plume, from the tailings pond, each of the pH zones, and a downgradient well near the property boundary. The samples are aligned along a geologic cross-section, which is depicted in Figure 6.3. The flow path is apparent not only from the pH contours, but also from hydrogeological data. The field parameters and major concentrations are listed in Table 6.1.

6.2.4   Selection of Modeling Code and Model Input

MINTEQA2, a geochemical modeling code published by US EPA (Allison *et al.*, 1991) was chosen for conducting speciation and solubility calculations. Program PRODEFA2, an interactive preprocessor for MINTEQA2, was used to prepare the input files, as described in §4.5.1. Special attention should be paid to the input of alkalinity data, which is discussed in Chapter 3, as well as in the MINTEQA2 manual on p. 20.

MINTEQA2 uses H_4SiO_4 as the silica component (basis species), whereas laboratories commonly report silica as SiO_2. If the units of concentration are moles per liter or moles per kilogram, no conversion factor is needed, but if units are in milliequivalents per liter, as they usually are, a concentration conversion factor of 1.6 is needed.[1]

---

[1]Calculated as follows:

$$\text{conc. of } H_4SiO_4 \text{ (mg L}^{-1}) = \text{conc. of } SiO_2 \text{ (mg L}^{-1}) \times \frac{\text{mol. wt } H_4SiO_4}{\text{mol. wt } SiO_2}$$

$$= \text{conc. of } SiO_2 \text{ (mg L}^{-1}) \times \frac{96.116}{60.085}$$

$$= \text{conc. of } SiO_2 \text{ (mg L}^{-1}) \times 1.60$$

```
12.00 MG/L   0.000   0.00000E-01
1 0 1 0 3 0 0 0 1 1 0 0 0
0   0   0
    330   0.000E-01   -7.40 y                    /H+1
     30   2.300E+02   -7.00 y                    /Al+3
    150   4.200E+02   -2.40 y                    /Ca+2
    281   9.260E+02    0.00 y                    /Fe+3
    500   2.780E+02   -2.64 y                    /Na+1
    732   8.100E+03   -2.35 y                    /SO4-2
    410   4.200E+01   -3.36 y                    /K+1
    460   7.000E+02   -3.06                      /Mg+2
    180   4.000E+02   -3.15 y                    /Cl-1
    770   1.600E+01   -4.23 y                    /H4SiO4
    270   1.000E-01   -4.98 y                    /F-1
    580   1.000E-02   -6.98 y                    /PO4-3
    140   2.458E+00  -12.07 y                    /Total CO3-2 alkali
    470   3.590E+01   -2.90                      /Mn+2


  3   1
    330      4.5000      0.0000                  /H+1
```

Table 6.2. MINTEQA2 input data for sample MW-86, produced by PRODEFA2.

For low pH water samples from the plume, concentrations of carbonate species were not measured, but they should be low. In modeling, we assumed that TS-3 and MW-86 have a small amount of dissolved carbonate. As an example of the MINTEQA2 input files, Table 6.2 shows the input file for sample MW-86, prepared by using PRODEFA2.

6.2.5   Geochemical Modeling

The overall aim is to use chemical analyses of the groundwater to determine what chemical reactions between the water and the host aquifers are probably occurring as the water moves along the aquifer. If the factors controlling the acidity and metal content are understood, predictions concerning them are easier to make. To do this we will:

- calculate the activities of all dissolved species in each sample;

- calculate the Saturation Indices (SI) of important minerals, and plot them on the cross-section along the flow path toward the northwest;

- observe a stepwise change in pH of the solutions, and test a buffering hypothesis with a titration model;

- plot the samples on activity–activity diagrams for additional insight.

6.2.6 Modeling Results

The calculated concentrations for selected species are shown in Table 6.3, and the Saturation Indices are listed in Table 6.4. These results on aqueous speciation are plotted in Figure 6.4, and show the aqueous speciation or species distribution along the cross-section A–A'.

The distribution of species varies along the flow path because the water chemistry changes drastically. The upgradient and hence background, uncontaminated water (MW-36) is a Ca^{2+}–HCO_3^- type water with a pH slightly higher than neutral, while the tailings fluid (TS-3) is predominantly SO_4^{2-}-rich, with high concentrations of dissolved metals Al, Fe, Mg, and Na. Contaminated water downgradient from the tailings pond has elevated SO_4^{2-}, but the metal concentrations have dropped significantly. The aqueous speciation is thus controlled by both solution pH and major ion concentrations.

Comments on Results for Specific Components

The following are a few of the more important speciation modeling results.

Al^{3+} The dominant species are Al^{3+}–SO_4^{2-} complexes in low pH, high SO_4^{2-} water (TS-3, MW-86) and $Al(OH)_3^{\circ}$ in low SO_4^{2-} and near neutral waters (MW-36, MW-15, MW-12, MW-14). It is wise to bear in mind that accurate analyses for dissolved Al are very difficult to perform. Because of its very low dissolved concentrations, particulate and colloidal particles containing Al can dominate an analysis, unless great care is taken. Driscoll and Postek (1996) note that " . . . because particulate minerals exhibit a continuous size distribution, no absolute distinction between dissolved and particulate forms can be made, and results show a strong dependence on filter pore size". In the absence of other errors, analyses for Al should be regarded as maximum possible values, from the point of view of geochemical modeling. It should also be noted that if an Al content is not reported (commonly the case), no conclusions at all can be reached about the saturation state of any aluminosilicate mineral.

Fe^{3+} Modeling results indicate the dominance of the polynuclear hydroxy-complexes $Fe_3(OH)_4^{5+}$ in low pH, high Fe^{3+} solutions (about $\geq 10^{-2}$ molal). See Stumm and Morgan (1996) for a discussion of these polynuclear complexes. Notably, the trimer is dominant over Fe^{3+}– SO_4^{2-} complexes. In near neutral solutions, the species $Fe(OH)_2^+$ is dominant.

6.2.7 Analysis of Mineral Saturation Indices

Figures 6.5–6.9 show the calculated Saturation Indices (SI) for carbonate, sulfate, iron, aluminum, and manganese minerals at Bear Creek. The calculation results show that a number of minerals are supersaturated in our samples. However, not all minerals that are indicated to be supersaturated are actually present at the site. Our discussion here is thus focused on how to interpret the calculated SI values, or to identify the mineral phases that are most likely relevant to the system.

Component	Species	MW-36	TS-3	MW-86	MW-15	MW-12	MW-14
Mn+2	Mn+2	82.1	52.7	58.6	71.3	72.8	75.7
	MnSO4 AQ	16.1	46.4	40.6	18.8	20.6	21.2
	MnHCO3 +	1.7	0	0	8.7	5.7	2.6
Al+3	Al+3	0	20	26.6	0	0	0
	Al(OH)2 +	2.9	0	0	22.8	15.9	1.5
	Al(OH)4 -	17.4	0	0	1.3	1.7	20.8
	Al(OH)3 AQ	79.7	0	0	70.4	79	77.7
	AlSO4 +	0	28.8	32.4	0	0	0
	Al(SO4)2 -	0	51.1	40.1	0	0	0
	AlF +2	0	0	0	3.1	1.6	0
Ca+2	Ca+2	80.6	49.4	55	72.8	72.8	74
	CaSO4 AQ	18.4	50.6	45	22.6	24.3	24.4
	CaHCO3 +	0	0	0	4.5	2.9	1.3
Fe+3	FeOH2 +	75.1	5.4	34.8	97	95.2	56.1
	FeOH3 AQ	19.3	0	0	2.8	4.4	27.3
	FeOH4 -	5.6	0	0	0	0	16.6
	FeSO4 +	0	10.1	1.7	0	0	0
	Fe(SO4)2 -	0	7.1	0	0	0	0
	Fe2(OH)2+4	0	7.4	2.7	0	0	0
	Fe3(OH)4+5	0	64.7	56	0	0	0
	FeOH +2	0	4.2	3.8	0	0	0
Na+1	Na+1	99	90.1	0	0	97.6	98.1
	NaSO4 -	0	9.9	0	0	2	1.7
SO4-2	SO4-2	79.6	51.1	66.6	65.7	68.6	70.5
	MgSO4 AQ	3.2	11.3	14.3	11.9	8.5	4.9
	CaSO4 AQ	16.4	2.3	5.6	21.3	21.3	24.4
	AlSO4 +	0	6.3	3.3	0	0	0
	Al(SO4)2 -	0	22.5	8.1	0	0	0
	FeSO4 +	0	2.1	0	0	0	0
	Fe(SO4)2 -	0	2.9	0	0	0	0
	NaSO4 -	0		1	0	1.5	0
K+1	K+1	98.8	87.8	0	0	97.6	97.9
	KSO4 -	1.2	12.2	0	0	2.4	2.1
Mg+2	Mg+2	82.2	52.7	0	0	73.8	75.8
	MgHCO3 +	1.2	0	0	0	4.6	2
	MgSO4 AQ	16.5	47.3	0	0	21.6	21.9
Cl-1	Cl-1	100	99.9	100	100	100	100
H4SiO4	H4SiO4	99.8	100			100	99.6
F-1	F-1	96.7	0	0	0	77.5	93.6
	MgF +	2	0	0	0	6.8	4
	CaF +	1.3	0	0	0	2.1	2.5
	AlF +2	0	99.4	0	0	4.2	0
	AlF2 +	0	0	0	0	5	0
	AlF3 AQ	0	0	0	0	4.2	0
PO4-3	H2PO4 -	21.1	9.3	0	0	42.5	8.3
	MgHPO4 AQ	7	1.5	0	0	10.7	10.5
	CaHPO4 AQ	23	0	0	0	17.2	33.9
	CaH2PO4 +	0	0	0	0	3.5	0
	CaPO4 -	1.3	0	0	0	0	3.8
	HPO4 -2	46.1	0	0	0	23.6	41.4
	FeH2PO4 +2	0	88.9	0	0	0	0
	MgH2PO4 +	0	0	0	0	2	0
	MgPO4 -	0	0	0	0	0	1.2
Ba+2	Ba+2	100	100	100	100	100	100
CO3-2	CaHCO3 +	1.3	0	0	2.1	2.1	2.6
	HCO3 -	91.4	0	2.3	64.3	73.4	92.4
	H2CO3 AQ	6.5	99.5	97.6	31.3	22.7	3
	MgHCO3 +	0	0	0	2.1	1.5	0

Table 6.3. Species distributions (%) in monitoring wells.

NAME	MW-36	TS-3	MW-86	MW-15	MW-12	MW-14
CA-NONTRONIT	24.75	26.057	26.94	27.875	27.09	28
MG-NONTRONIT	24.341	25.882	26.718	27.543	26.733	27.592
K-NONTRONITE	18.735	20.224	21.028	21.787	20.976	21.617
NA-NONTRONIT	17.949	19.527	20.332	21.172	20.429	21.054
MUSCOVITE	5.195	0.98	4.081	11.548	11.249	10.633
KAOLINITE	3.856	3.154	4.888	8.729	8.476	7.939
PYROPHYLLITE	3.611	5.055	5.896	9.576	9.288	8.772
DIASPORE	1.973	0.524	1.949	3.921	3.84	3.533
HALLOYSITE	0.474	-0.215	1.471	5.324	5.059	4.535
BOEHMITE	0.179	-1.261	0.127	2.108	2.018	1.721
GIBBSITE (C)	0.123	-1.334	0.112	2.079	2.004	1.692
LOW ALBITE	-1.499	-2.529	-2.352	1.434	1.484	1.456
ANALCIME	-1.531	-3.668	-2.914	0.92	1.02	0.949
ALOH3(A)	-1.595	-3.041	-1.64	0.339	0.252	-0.048
LAUMONTITE	-1.657	-7.56	-5.627	2.63	2.563	4.157
ANALBITE	-2.479	-3.503	-3.353	0.44	0.483	0.462
Al2O3	-4.034	-7.06	-3.689	0.122	0.093	-0.653
KALSILITE	-4.358	-7.862	-6.544	-2.905	-2.964	-3.03
ANORTHITE	-4.675	-12.751	-9.771	-1.397	-1.39	0.143
NEPHELINE	-5.061	-8.286	-7.01	-3.116	-2.979	-3.08
WAIRAKITE	-6.39	-12.255	-10.459	-2.169	-2.271	-0.643
SPINEL	-8.231	-17.091	-13.33	-5.501	-5.559	-4.395
CRYOLITE	-19.168	-34.164	-32.062	-14.299	-12.872	-18.311
GEHLENITE	-20.916	-37.464	-32.709	-19.742	-19.436	-15.863
JAROSITE K	4.443	16.237	16.952	9.123	7.446	4.293
JAROSITE NA	1.504	13.579	14.246	6.673	5.19	2.004
BARITE	0.787	1.466	1.428	0.967	1.018	0.989
GYPSUM	-0.553	0.217	0.296	0.164	0.127	0.024
ANHYDRITE	-0.867	-0.081	-0.047	-0.169	-0.218	-0.31
ALUNITE	-1.123	7.633	10.206	8.46	7.736	2.625
AL4(OH)10SO4	-2.291	-0.105	5.093	8.147	7.694	4.126
EPSOMITE	-3.848	-1.691	-1.869	-2.664	-2.84	-3.251
ALOHSO4	-4.779	1.956	2.092	-0.574	-0.983	-3.433
MIRABILITE	-6.361	-4.021	-4.207	-4.957	-4.691	-6.564
THENARDITE	-7.792	-5.378	-5.783	-6.488	-6.274	-8.098
MNSO4	-11.757	-8.647	-8.922	-11.25	-11.948	-11.056
ALUM K	-16.162	-4.41	-5.163	-10.722	-11.407	-15.779
FE2(SO4)3	-43.244	-19.405	-22.302	-36.166	-38.108	-43.686
HEMATITE	17.507	16.527	18.581	18.356	17.672	18.251
MAGHEMITE	7.898	6.836	9.217	8.91	8.308	8.805
MAG-FERRITE	7.246	0.292	3.3	6.951	6.381	8.727
FEOH)2.7CL.3	6.999	7.924	8.867	8.114	7.715	7.523
GOETHITE	6.273	5.78	6.817	6.703	6.363	6.65
LEPIDOCROCIT	5.771	5.239	6.43	6.277	5.976	6.224
JAROSITE K	4.443	16.237	16.952	9.123	7.446	4.293
FERRIHYDRITE	2.251	1.716	2.909	2.756	2.455	2.704
JAROSITE H	-2.841	12.198	12.09	2.637	0.798	-2.463
CALCITE	-0.026	-7.644	-6.04	0.374	0.309	0.889
ARAGONITE	-0.2	-7.813	-6.226	0.193	0.123	0.708
DOLOMITE	-0.761	-14.578	-11.719	0.466	0.176	1.048
MAGNESITE	-1.228	-7.429	-6.166	-0.398	-0.621	-0.332
RHODOCHROSIT	-1.379	-6.702	-5.289	-1.11	-1.794	-0.26
NESQUEHONITE	-3.627	-9.836	-8.565	-2.796	-3.018	-2.729
WITHERITE	-3.796	-11.499	-10.055	-3.956	-3.944	-3.279
HUNTITE	-6.438	-32.643	-27.312	-3.577	-4.328	-2.864
ARTINITE	-8.009	-20.228	-17.866	-8.255	-8.328	-6.306
NATRON	-9.288	-15.338	-14.013	-8.211	-7.975	-9.161
THERMONATR	-11.195	-17.174	-16.057	-10.212	-10.027	-11.165
HYDRMAGNESIT	-17.542	-48.352	-42.253	-15.323	-16.079	-13.175

Table 6.4. Calculated Saturation Indices for selected minerals.

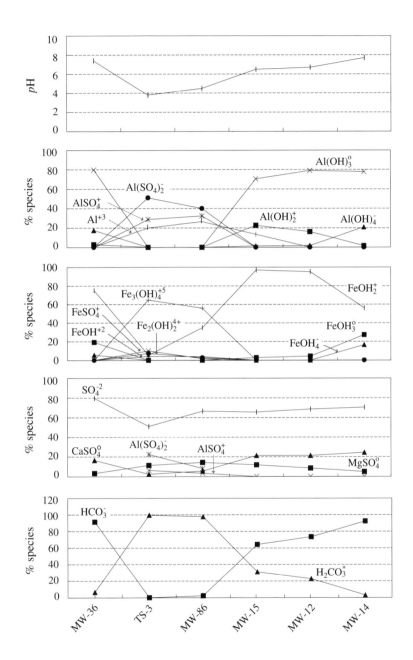

Figure 6.4. Changes of species distribution in groundwater samples from cross-section A–A′ from Figure 6.3.

   The decision as to which mineral phases are significant in a system based on modeling results is a common problem for geochemists. Speciation–solubility modeling only can point out the possibilities of solubility control. Positive identification of the solubility limiting phases can only be accomplished by microscopic or X-ray analysis of soils and aquifer matrix. As we pointed out earlier, mineralogical work is often not a part of groundwater studies. One reason for this is that it is difficult work, given the often extremely fine-grained or amorphous phases and solid solution effects (Jambor and Blowes, 1994). In these cases, the modeler's geological knowledge becomes significant. Modelers can also refer to mineralogical studies with similar geological and environmental conditions.

   Given accurate analyses and thermodynamic data, Saturation Indices can only show which minerals could conceivably precipitate from the solution (SI>0) (and could therefore be present in the system), and which minerals absolutely could not (SI<0). There are several reasons why many of the minerals which are calculated to be supersaturated may not actually be present in the system.

- Minerals which are indicated to either dissolve or precipitate may actually not do so because of kinetic constraints. This applies especially to precipitation of complex structures which may form quite easily at higher temperatures. Quartz is almost universally supersaturated in surface waters, and if SiO_2 does precipitate, it will be in some other form, such as amorphous silica.

- Calculated SI are for pure "end-member" minerals, but the actual minerals in environmental settings can be solid solutions with complicated compositions. Alpers *et al.* (1989) found that, for jarosite, errors of more than one order of magnitude can be introduced by ignoring solid solutions.

- The magnitude of the SI is often of little importance. Bethke (1996) and Wolery (1992) point out that the SI value depends not only on the concentrations involved, but on the formula of the mineral. For example, if we were to write the formula of quartz as Si_2O_4 instead of SiO_2, its Saturation Index would be doubled. The choice of large formula units for clays and zeolites explains why these minerals often occur at the top of the supersaturation list (Bethke, 1996, p. 91).

- Errors in analyses may be significant. We have already mentioned the problem with particulate and colloidal aluminum, and the same point could be made about silica and many other elements present in small amounts.

- The reliability of much of the thermodynamic data is always questionable. The data are obtained in a number of different ways, some more reliable than others. For solid phases, the nature of the solid phases involved in the determination of the data may be different from those in the system from the point of view of grain size, crystallinity, defect structure, and composition.

- Of course, even if the calculated SI is accurate, e.g., a given mineral is in fact present and is dissolving (SI < 0), or is in fact precipitating (SI > 0), the SI tells us nothing at all about the rate at which this is happening, nor about the quantities of minerals involved.

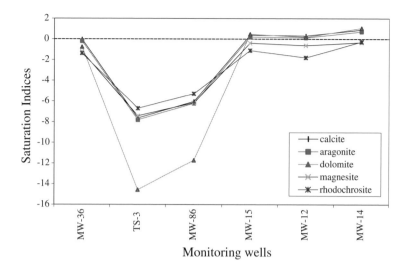

Figure 6.5. Saturation Indices for carbonate minerals from MINTEQA2.

On the other hand, many investigations have shown that Saturation Indices are a reasonably good guide to certain phases, such as some carbonate and sulfate minerals. Saturation Indices are also quite useful in reaction path modeling (Chapter 8) and inverse mass balance modeling (Chapter 9) in the sense that they indicate (within the accuracy of the data) which processes are possible (e.g., precipitation of a phase having an SI > 0), and which are impossible (e.g., dissolution of a phase having an SI > 0).

Carbonate Minerals   (Figure 6.5) SI values of calcite for the upgradient well MW-36, and wells downgradient from the plume MW-15 and MW-12 water are within 0 ± 0.5, a range well within the uncertainties propagated from errors associated with sampling, laboratory analyses, and thermodynamic properties. MW-14 shows an SI of 0.9. Equilibrium with calcite at the site is easy to be inferred because we know that the N sand and alluvium contain about 2% calcite. Calculated SI values for aragonite are within 0 ± 0.2, but its presence is unlikely. Modeling results also show supersaturation or equilibrium with dolomite in waters downgradient from the plume, but dolomite is undersaturated in the upgradient water (MW-36). Whether or not dolomite is present in the sandstone is not independently confirmed, but precipitation of ordered dolomite would be rather unusual under surface conditions. In all water samples, magnesite ($MgCO_3$) is undersaturated.

Sulfate Minerals   (Figure 6.6) Fifteen sulfate minerals are present in the thermo-dynamic database. The calculations show that many are supersaturated in our water samples. Several members of the alunite–jarosite group are grossly supersaturated, but alunite and jarosite are known to be supersaturated without precipitating in some acidic surface waters (Alpers *et al.*, 1994). Aluminum hydroxy sulfates, $Al(OH)SO_4$ (jurbanite), and $Al_4(OH)_{10}SO_4$ (basaluminite), are also supersaturated at some points

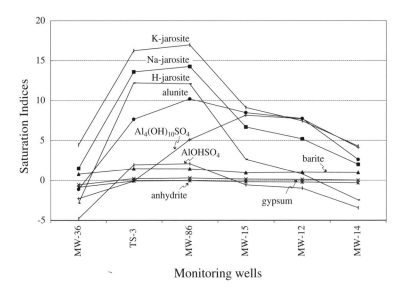

Figure 6.6. Saturation Indices for sulfate minerals from MINTEQA2.

in the cross-section. These minerals have been found in mine drainage settings (Alpers *et al.*, 1994). Stollenwerk (1991) found that using a modified jurbanite solubility product fitted well with the Al results in column experiments on Pinal Creek acid water at $pH < 4.7$. On the other hand, mineralogical studies of acid mine drainage sites indicate that gypsum is the most common sulfate phase controlling sulfate concentrations in such environments (Davis *et al.*, 1991; Jambor and Blowes, 1994).

Iron Minerals   (Figure 6.7) The calculations show that a number of iron minerals are supersaturated in the fluids. Besides jarosite, the water samples are also supersaturated with goethite and hematite, both of which have slow growth kinetics at surface temperatures. Jambor and Blowes (1994) argue that goethite with various crystallinities is the iron phase that commonly controls iron solubilities. Our calculations show that SI values for ferrihydrite, defined in the MINTEQA2 database as an amorphous $Fe(OH)_3$, are closer to zero than are those for well crystallized goethite. Its SI values are about two log units above zero. These SI, however, are a function of the log K values in the database. The equilibrium constants for amorphous iron hydroxide or oxide vary several orders of magnitude (Cornell and Schwertmann, 1996; Stumm and Morgan, 1981, p. 240). MINTEQA2 uses a value of K_{sp} of -4.89 from Nordstrom *et al.* (1990), but the NBS value reported by Wagman *et al.* (1982) is -5.66.

Aluminum Minerals   (Figure 6.8) Similarly, a large number of aluminum minerals are supersaturated in the sampled water. The sandstone and alluvium materials contain aluminum silicate minerals and clays, but these minerals are unlikely to be at equilibrium in groundwater due to kinetic inhibitions. The calculated SI for clays are again for end-members. In our samples, gibbsite or an amorphous analog appears to be the

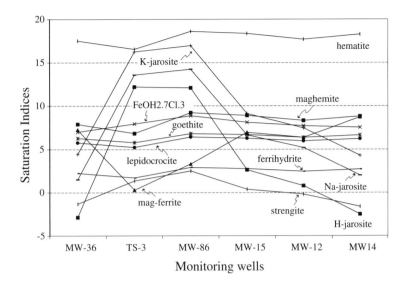

Figure 6.7. Saturation Indices for iron-bearing minerals from MINTEQA2.

closest to equilibrium. It is fairly common to find that the Al^{3+} content of soil waters is apparently controlled by the solubility of a gibbsite-like mineral, although gibbsite is rarely identified mineralogically (Hendershot *et al.*, 1996). This may be due to an amorphous gibbsite-like phase, or to surface adsorption effects. Stollenwerk (1991) found that using $Al(OH)_3(a)$ from MINTEQA2 and a modified log K for jurbanite successfully explained solubility control of his experimental data. Jurbanite controls Al solubility at $pH < 4.7$ and $Al(OH)_3(a)$ controls solubility at $pH > 4.7$.

Mn Minerals   (Figure 6.9) All water samples are undersaturated with respect to all Mn^{2+}-containing mineral phases available in the MINTEQA2 thermodynamic database. However, along the flow path the Mn concentrations drop from highly contaminated to only slightly contaminated groundwater, so that Mn must have been precipitated in some form. The most likely precipitate in surficial environments is MnO_2 (Hem, 1985), but the small amount of Mn involved may simply have been adsorbed on mineral surfaces.

6.2.8   Activity–Activity Diagrams

Additional insight into the reactions involved is obtained by plotting the solution compositions on appropriate activity–activity diagrams (Figure 6.10). This is most conveniently done with programs in the The Geochemist's Workbench™ group (Bethke, 1994), because they are among the few geochemical modeling programs that can produce diagrams. However, the same diagrams can be produced from the output of other programs with a little more effort.

Notice in Figure 6.10 how the upgradient background (MW-36) fluid plots higher in the diagram, and the more contaminated fluids (MW-86, TS-3) plot lower, because

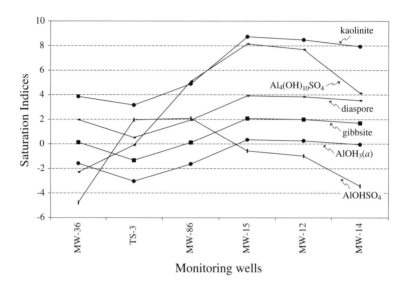

Figure 6.8. Saturation Indices for aluminum-bearing minerals from MINTEQA2.

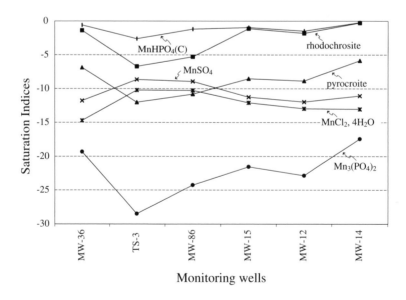

Figure 6.9. Saturation Indices for Mn-bearing minerals from MINTEQA2.

"Downstream Equilibrium Conditions"

In analysis of Saturation Indices, it is tempting to infer that minerals are present in the regions where SI $\geq$ 0. However, Walsh *et al.* (1984) noted that fluid in the downstream direction of a set of minerals is saturated with respect to this set of minerals even though the solids themselves are not present in the downstream region. They termed this condition the "downstream equilibrium condition". It is a result of physics and thermodynamics, under the imposed local equilibrium assumption (Lichtner, 1996; Walsh, 1983).

If we examine the SI values at the Bear Creek Uranium site, we see that the uncontaminated background groundwater is undersaturated with respect to gypsum (SI $<$ 0, MW-36 in Figure 6.6). The acidic groundwater (MW-86) in the Al(OH)$_3$(*a*) buffer zone has a SI $\approx$ 0, and, according to our geochemical evolution model, there should be gypsum present in the plume. The calcite dissolution–gypsum precipitation front should be located somewhere between MW-86 and MW-15, judging from the pH of the samples (4.5 and 6.5, respectively). However, samples downstream from the gypsum precipitation zone, MW–15, MW-12, and MW-14 all have SI $\approx$ 0. This distribution of SI values may indicate the "downstream equilibrium condition", rather than the presence of gypsum in these areas, which we know is not present in the alluvium. We note that gypsum precipitation is fast and that the local equilibrium assumption is probably valid.

---

larger values of a_{H^+} (lower pH) produce lower values of the ratio a_{K^+}/a_{H^+}. The more contaminated fluids also have a greater silica content and so plot more to the right.

Figure 6.10 shows solution compositions (at least the log a_{K^+}/a_{H^+} and log $a_{SiO_2(aq)}$ aspects of the compositions) calculated by MINTEQA2 plotted on mineral boundaries calculated by The Geochemist's Workbench™. It is desirable not to mix databases in this way. To show the differences between these databases, Figure 6.11 shows the mineral boundaries calculated from both MINTEQA2 and The Geochemist's Workbench™. These differences are typical of those one can expect by changing databases.

6.2.9 Geochemical Evolution Along A Flow Path

The speciation–solubility modeling results, combined with other numerous groundwater data that are presented here, and knowledge of water–rock interactions, help to conceptualize geochemical evolution along the flow path. As discussed earlier, groundwater monitoring data and geochemical modeling have delineated four geochemically distinct zones. The existence of geochemical zones is mainly supported by the observation that the pH distribution is discontinuous. Distinct pH zones are also observed at many other acid mine drainage environments (Morin *et al.*, 1988), and is explained by

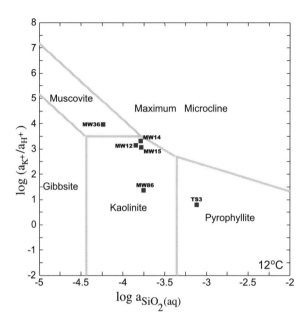

Figure 6.10. Projection of water sample compositions (calculated by MINTEQA2) onto the $\log a_{K^+}/a_{H^+}$ vs. $\log a_{SiO_2(aq)}$ diagram at 12°C.

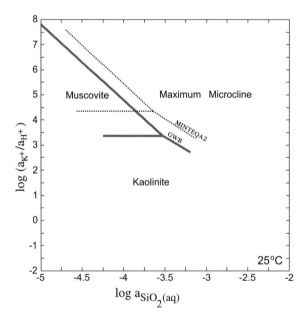

Figure 6.11. The same curves as drawn by The Geochemist's Workbench™ and from the data in MINTEQA2, at 25°C.

the successive buffering process discussed in §8.4.

Based on this conceptual model and site data, we can decipher the groundwater geochemical evolution along the flow path in cross-section A–A$'$ (Figure 6.3). Groundwater at MW-14 has not yet been impacted by acid as indicated by its relatively high pH, but higher concentrations of metals and Cl^- and SO_4^{2-} have already reached the area (see Table 6.1 and compare concentrations). Groundwater at MW-15 and MW-12 are at the first stage of sulfuric acid intrusion. Calcite in the aquifer reacts with the infiltrating low pH water and buffers the fluid pH to above 6.5:

$$CaCO_3 + H^+ = Ca^{2+} + HCO_3^-$$

As long as calcite is present, the pH will be maintained at about 6.5 while calcite in the aquifer matrix is being consumed. This reaction is evident from the higher Ca^{2+} and HCO_3^- concentrations compared with those in background water as represented by MW-36.

The dissolution of calcite also results in higher calcium concentration and leads to gypsum precipitation:

$$Ca^{2+} + SO_4^{2-} + 2H_2O = CaSO_4 \cdot 2H_2O$$

As a result, sulfate concentrations are reduced, compared with those in upgradient water.

Groundwater at MW-86 is at the second geochemical phase when calcite in the aquifer is depleted. The fluid pH of the groundwater is reduced to about 4.5 and is buffered by amorphous aluminum hydroxide or kaolinite.

$$Al^{3+} + 3H_2O = Al(OH)_3(aq) + 3H^+$$

Precipitation reduces the aluminum and iron concentrations in the fluid to a few milligrams per liter. The fact that there are few observed groundwater pH values between 6.5 and 4.5 supports the hypothesis that solid phases buffer fluid pH at the site.

The pH of tailings fluids (TS-3) appears to be controlled by iron hydroxide:

$$Fe^{3+} + 3H_2O = Fe(OH)_3(aq) + 3H^+$$

The fluid is undersaturated with respect to gibbsite or amorphous Al hydroxide. As more tailings fluids seep into the shallow aquifer, the reaction zones will shift downgradient.

6.2.10   Comments on the Bear Creek Site

The speciation and solubility modeling of six water analyses gives us:

- six snapshots of the speciation of cations and anions in the aqueous phase and the equilibrium state with respect to minerals in a dynamic system;

- identifcation of the major chemical reactions that control the acid plume migration;

- a conceptual model of geochemical evolution along the flow path at the site;

- identified reactions that can be significant to heavy metal and radionuclide retardation.

The modeling results show that a large amount of iron and aluminum hydroxide will be precipitated. Coprecipitation with iron and aluminum hydroxide is commonly found and is a significant natural attenuation process (Herbert, 1996; National Research Council, 2000; Zhu *et al.*, 1993).

In terms of answering the ultimate question "what concentrations will be at the point of compliance over the next 200 years?", the speciation–solubility calculations appear to provide only tentative results. The Saturation Indices only provide information about the equilibrium state, but do not consider masses or the processes that are needed for the aquifer to neutralize the acid plume. The equilibrium modeling also provides only a snapshot of the state of equilibrium in a dynamic system, in which the controlling reactions vary over time and space. Besides geochemical reactions, the potential migration of acid and metals is also influenced by factors such as advection and hydrodynamic dispersion. Does that mean that geochemical modeling is not useful? Not at all. A good analog of our speciation–solubility results is the equal-potentiometric surfaces prepared for a groundwater flow model. The potentiometric surfaces themselves do not provide answers to all practical questions regarding groundwater flow, but they do provide a basis for constructing a conceptual flow model. While the potentiometric surfaces indicate the possible directions of the groundwater flow, Saturation Indices point out the possible directions of chemical reactions. Similarly, the equilibrium reactions between aquifer matrix and groundwater characterized from speciation–solubility modeling at a site provide a geochemical basis for any meaningful contaminant transport models.

6.3   Applications to Bioavailability and Risk Assessment Studies

Davis *et al.* (1992, 1993, 1996) and Hemphill *et al.* (1991) contend that geochemical factors control the bioavailability of lead (Pb) and arsenic (As) and that human risk assessments based on the assumption that soil Pb and As are equally bioavailable as soluble salts are flawed. When soluble salts, for example $Pb(OAc)_2$, are used in toxicological studies, the amount of Pb in body fluids is controlled by supplied salt mass. For mine-waste impacted soils, not all Pb and As are soluble. In fact, the solubility may be much less and is controlled by geochemical factors that include the dominant metal-bearing phases, encapsulation, and rinding by less soluble phases, and particle sizes. They reasoned that soil Pb and As should be less bioavailable than soluble salts because Pb and As in the mining wastes are less soluble. Bioavailability is expressed as the percentage of ingested materials absorbed through the intestinal wall into the systemic blood system. The mineralogical controls can help to explain the variable blood Pb levels in different settings (urban, smelter, and mining waste) with similar soil Pb levels.

Davis *et al.* (1992, 1993) first studied the mineralogy of the mining waste impacted soil in the Butte, Montana, mining district, and then performed speciation–solubility calculations. They reasoned that dissolution of As and Pb under the acidic stomach

Solid phase	Solubility	Distribution %
Galena (PbS)	69 mg/L	72 Pb^{2+}
		28 $PbCl^+$
Anglesite (PbSO$_4$)	69 mg L^{-1}	66 Pb^{2+}
		26 $PbCl^+$
		4 $PbCl_2^o$

Table 6.5. Solubility control of lead concentrations in stomach fluids (pH 2.0, 0.01 M HCl, and Eh +200 mV. Calculated using MINTEQA2 (Davis *et al.*, 1992).

conditions is the necessary precursor for absorption of these metals by the small intestine. Therefore, solubility of Pb and As from equilibrium modeling is intended to provide an upper bound of bioavailable Pb and As in the gastrointestinal (GI) tract.

Davis *et al.* (1992, 1993) used MINTEQA2 for their solubility calculations. The stomach fluid is simulated by using a 0.01 M HCl solution at a pH of 2.0 under an oxidizing condition of Eh of 200 mV, and the small intestinal fluid by a pH of 7.0 and Cl^- of 10^{-7} M. The minerals anglesite (PbSO$_4$) and enargite (Cu$_3$AsS$_4$) are believed to be the most likely phases controlling Pb and As solubility. The calculated solubility of anglesite is 69 mg L^{-1} in stomach fluid and 37 mg L^{-1} in the small intestine. Addition of 10^{-4} M each of acetate and citrate acids increase the solubility of Pb to 61 mg L^{-1} in the small intestine because of the formation of organic acid–Pb complexation. Addition of phosphate and calcium caused the precipitation of lead phosphate phases and a lower solubility control. Table 6.5 lists the calculated solubility and aqueous species distribution.

For As solubility calculations, the modeling generated less useful information. Enargite is found to be quite soluble in stomach fluids. Davis *et al.* (1992) believe that dissolution kinetics or inhibition from reaction with GI fluids by encapsulation with less soluble phases (e.g. jarosite) cause the observed discrepancy between equilibrium calculations and experiments.

Comparing the modeling results with *in vivo* experiments using female New Zealand white rabbits shows that the *in vivo* Pb concentration is lower than the up-bound of soluble Pb calculated from anglesite solubility in stomach fluid. The calculated Pb solubility is 69 mg L^{-1}, whereas the *in vivo* Pb concentration in the rabbit stomach fluid is 8.5 mg L^{-1}. Ruby *et al.* (1992) attributed the lower experimental Pb concentration to dissolution kinetics of anglesite and a short transit time in the GI tract (6 hours).

While Davis and his colleagues illustrated the significance of soil metal speciation in risk assessment, Morrison *et al.* (1989) pointed out that the toxicity of metals is related to the forms in which they exist in the aqueous phase. This is because the interaction of metals with intracellular compartments is highly dependent on chemical speciation. Some species may be able to bind chemically with extracellular proteins and other biological molecules, some may adsorb onto cell walls, and others may diffuse through cell membranes. Consequently, toxicity is more related to the concentrations of metals in a particular species, than to the total concentrations. Geochemical modeling can be used to calculate the equilibrium species distribution for measured total metal concentrations. Morrison *et al.* (1989) also pointed out that the conclusions about toxic species in the literature are based on calculated equilibrium speciation distributions. Therefore, it is the relevant metal species, not the total concentration, that should be used in risk assessment.

Examples showing that metal speciation is important to metal toxicity include arsenic, copper, selenium, and chromium. While ionic copper (Cu^{2+}) and $CuCl_2^0$ are highly toxic, $CuCO_3^0$ and Cu-EDTA have low toxicity (Morrison *et al.*, 1989). Toxicity tests show that As(III) is about 50 times more toxic than As(VI). Trivalent chromium is much less toxic than hexavalent chromium, probably because Cr(VI) is much smaller and the chemical structure of chromate is similar to sulfate. A special channel already exists in biomembranes for sulfate transport. While modeling metal speciation is not always possible, and redox equilibrium is not achieved in all natural waters, geochemical modeling of equilibrium species distribution remains one of the methods of discerning metal speciation.

6.4   Interpretations of Column Experiments

Li *et al.* (1996) reported the use of speciation–solubility modeling to interpret the effluent chemistry in column experiments. Noranda Mining and Exploration, Inc. was in the early stage of opening an open pit copper mine at Mines Gaspé in eastern Quebec, Canada. Approximately 20 million tons of oxide ore were to be dissolved with sulfuric acid using a heap leaching operation. After the conclusion of the heap leach operation, the leachate from the spent heap, as a result of rainfall percolation, must meet Canadian government discharge requirements (<0.5 mg L^{-1} for Cu). One decommission option was to rinse the spent heap with clean water and neutralize the acid leachate with limestone or lime.

A large, "pilot scale" column measuring 671 cm in height and 66 cm in diameter was used to study this decommissioning option. Broken but uncrushed ores of about 6 inch size were placed in a column, and rinsing studies commenced after the completion of the copper extraction experiments. Clean water was fed into the top of the column, and leachate was collected weekly at the bottom of the Gaspé Large Column 2.

Figures 6.12, 6.13, and 6.14 show the breakthrough curves for pH and major cations and anions from Gaspé Large Column 2. Rinsing volumes were normalized by dividing the total cumulative volumes of water passed through the column by the dry ore weight in that column. Figures 6.15 and 6.16 show the Saturation Indices (SI) calculated from MINTEQA2 using measured pH and ion concentrations.

The authors concluded that the leachate reached equilibrium with silica and gypsum because the SI values of these two minerals are constant and close to zero despite the orders of magnitude of concentration differences in sulfate. Solubility control by gypsum also explains the calcium concentration increase at the early washing stage. Calcium concentrations were depressed at first because of the high sulfate concentrations. With the sharp decrease of sulfate concentrations, calcium concentrations increased in response to the decrease in sulfate in the solution.

The solubility control of iron concentrations is evident by the constant SI of goethite

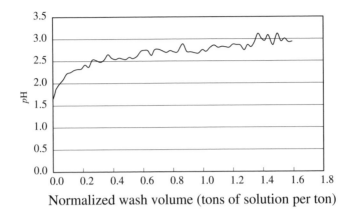

Figure 6.12. pH responses to rinsing in Gaspé Large Column 2.

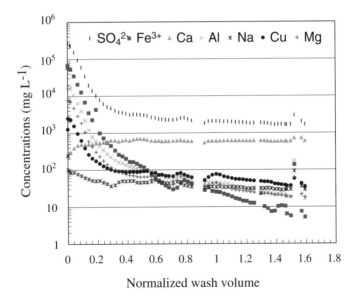

Figure 6.13. Ion concentrations of leachate in Gaspé Large Column 2.

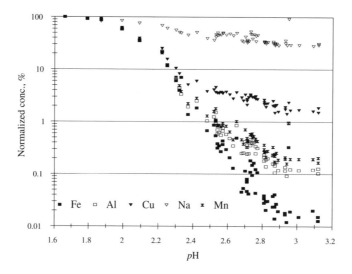

Figure 6.14. Decrease of normalized concentrations as a function of pH.

in leachate with varying Fe^{3+} concentrations. Li *et al.* believe that an amorphous iron hydroxide such as ferrihydrite controls the iron solubility. In fact, we see a pH buffer effect by $Fe(OH)_3$, similar to the uranium mill tailings site discussed in §6.2. Here, the pH buffer occurs at about 0.25 wash volume or a pH of about 2.7. We also observe from Figure 6.14 that Fe^{3+} concentrations decrease by three orders of magnitude with an increase of one pH unit.

The interpretation by Li *et al.* (1996) was that Al was controlled by $Al(OH)SO_4$ solubility at low pH but that this phase was depleted as the washing proceeded. The authors also interpreted the wide range of SI values for jarosites to be a result of the absence of jarosite in the column.

From these column experiments, the authors projected that six wash volumes were needed to rinse the spent heap to meet the government discharge standards. They concluded that relying on rainfall rinsing would take 338 years, but spraying with artificial precipitation using the heap leach irrigation system would be more economical. They estimated that it would cost $5.2 million (Canadian) for this decommission option. Presumably, results from this work were also used in the mining permit application as part of the reclamation plan.

It should be noted that the leachate from the column experiments contains hundreds of thousands of milligrams per liter of dissolved solids, resulting in high ionic strength for the solutions. MINTEQA2 uses the Davies equation and can only accurately simulate solutions with an ionic strength of approximately $\leq 0.2\,m$ (§3.4.2). Therefore, the calculated SI values discussed above bear the errors associated with activity coefficients. It would therefore be useful to recalculate the results of at least one experiment using Pitzer activity coefficients (p. 41), to see if the errors are significant.

Besides geochemistry, transport processes also exert significant controls on effluent concentrations. The dual porosity model (Decker, 1996; Dixon *et al.*, 1993) for

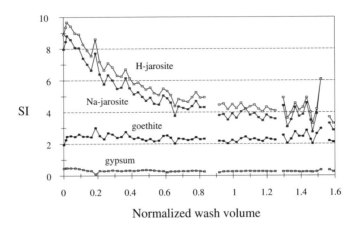

Figure 6.15. Saturation Indices for leachate for Gaspé Large Column 2.

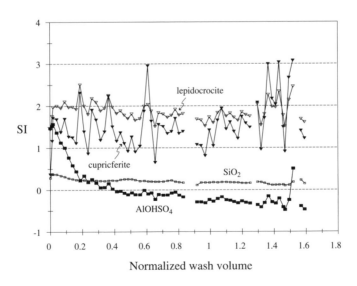

Figure 6.16. Saturation Indices for leachate for Gaspé Large Column 2.

columns packed with spent heap consider macropores where fluid flows preferably and microporosity where fluid flow is stagnant. During initial washing, the leachate residues in the spent heap are displaced by advective transport. Then, the concentrations in the effluent are controlled by matrix diffusion. In reality, reaction kinetics may also come into play in the fast flow and stagnant regions. Speciation and solubility calculations only concern the "state variable", not the quantity of mass or the transport processes. However, they serve as necessary building blocks for an eventual quantitative model.

A similar modeling study was conducted by Stollenwerk (1994) to interpret a column experiment on the acidic water and alluvium sediments interactions for Miami Wash and Pinal Creek, Arizona. He used MINTEQA2 to simulate the equilibrium states in column effluents.

7

Modeling Surface Adsorption

7.1 Introduction

Thermodynamics, as normally defined and as presented in Chapter 3, contains no reference to surfaces. Phases, such as clay minerals and water, are assumed to exist and to equilibrate, based on their bulk properties only. However, in reality, phases interact with each other along interfaces or surfaces, the properties of which are necessarily different from the bulk properties of the phases.

On the surfaces of a phase, the normal environment of each atom is changed, and the atoms are forced to interact with atoms of a different sort, provided by the adjacent phase. In most solids and liquids, bonding is effected by electrical effects–electron transfer, electron sharing, polarization effects, and so on. In the middle of a phase there is a net charge balance, but this is disrupted at surfaces, where the three-dimensional structure is broken. Surfaces are zones where atoms are left with unsatisfied bonds, and therefore surfaces are electrically charged. These charges are accommodated somehow by the adjacent phase, and the case of most interest to geochemical modelers is the case in which one phase is a solid, and the other is water.

7.1.1 The Solid–Water Interface

This interface is quite complex, and is the subject of much active research. We present here the minimum amount of information required to understand what most geochemical modeling programs provide in the way of adsorption modeling at the present time. The capabilities present in widely used modeling programs are of two types – ion-exchange and surface complexation modeling. These represent both the historical development of the subject, and, to some extent, our understanding of two important ways in which ions adsorb to mineral surfaces.

Many minerals in aqueous solutions in the normal range of pH values have a negatively charged surface, which becomes neutralized by attracting positively charged particles (cations). These attracted ions may reach the solid surface and actually bond to the surface ions (inner- and outer-sphere complexes) in the *Stern layer* (Figure 7.1),

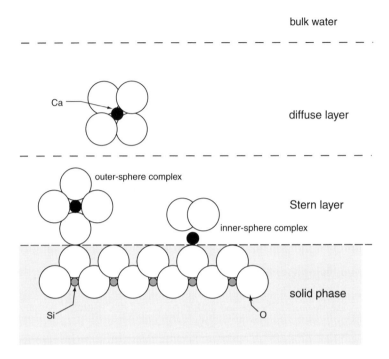

Figure 7.1. A schematic representation of a mineral surface. Hydrated aqueous ions may bond to the surface in the Stern layer, or reside close to the surface in the diffuse layer. Beyond that, "bulk water" represents water unaffected by the presence of the surface. Not to scale. (Modified from Dzombak and Hudson, 1995, Fig. 3.)

or they may simply reside close to the surface in the *diffuse layer*. The Stern layer and the diffuse layer together constitute the "electrical double layer". These two modes of adsorption can be viewed, albeit approximately, as the basis of the two kinds of adsorption modeling.

7.2   Ion-exchange

7.2.1   Cation-exchange Capacity

The negative charge on a clay mineral surface in a solution containing NaCl will be counteracted or neutralized by Na^+ ions in both the layers mentioned, but mostly in the diffuse layer. If the solution is then replaced by one containing mostly $CaCl_2$ instead of NaCl, the adsorbed Na^+ ions will be quickly replaced by (half as many) Ca^{2+} ions in a process called *ion-exchange*. More generally, whatever ions are adsorbed to the surfaces of a mineral or soil sample, they can be replaced by (exchanged for) another set of ions. This process has been more or less standardized in a batch process, in which a sample is eluted with a solution containing a standard ion, often NH_4^+, and the number of moles of ions displaced from the sample by the NH_4^+ is measured. This quantity, the total exchangeable equivalents of cationic charge under specified conditions, is called the Cation-exchange Capacity, or CEC. It is normally measured in milliequivalents per 100 g of sample. Anion exchange can be measured in the same way, using a replacing anion, but this is usually a less important quantity.

Although "ion-exchange" is most often thought of in terms of surface as described above, it can also apply to exchange of ions from within a phase, as in the use of zeolites in water softening, or sodium for potassium exchange when K-feldspar is altered to albite at high temperatures.

7.2.2 Exchange Reactions

In the case of calcium ions replacing sodium ions, the exchange process can be written

$$Ca^{2+} + 2\,NaX = CaX_2 + 2\,Na^+ \tag{7.1}$$

In this equation, X represents an "exchangeable site", which may actually be a site in a mineral or on a mineral surface. However, usually it is not known just how or where the ions are adsorbed, and X^- is then simply a component representing a unit charge anywhere in or close to the solid phase (Dzombak and Hudson, 1995). In either case, though, some manipulation of the CEC allows us to assign a value K_X to the ratio

$$\frac{[Na^+]^2[CaX_2]}{[Ca^{2+}][NaX]^2} = K_X \tag{7.2}$$

where [ ] denotes concentration. Using X^- as such a component, exchange equations such as (7.1) can be broken into two "half-reactions", just as redox reactions are broken into two half-cell reactions (§3.8.2). In this case, these would be

$$Na^+ + X^- = NaX \tag{7.3}$$

and

$$Ca^{2+} + 2\,X^- = CaX_2 \tag{7.4}$$

Subtracting twice Equation (7.3) from (7.4) then gives Equation (7.1). Using a suitable convention, K values can be assigned to each of these, such that K_X is obtained for the complete reaction, and it is in this "half-reaction" form that ion-exchange data is stored in databases.

There are several complications in the use of this concept.

- The measured CEC is highly sample specific, so that use of generalized ion-exchange "constants" such as K_X for conditions involving various solids is of dubious validity.

- Values of K_X are normally obtained from simple laboratory experiments involving only two cations. How to extrapolate to natural multi-component conditions is problematic.

- The concentrations of the solid phase sites (X) can be quantified, but their activities cannot, because activity coefficients are not known. Activities are therefore generally taken to be the same as the concentrations, although some rather arbitrary scheme can be used to calculate activity coefficients.

- The concentrations themselves can be calculated in different ways, depending on whether the fraction of sites occupied by an ion is given on a molar or equivalent basis (Appelo and Postma, 1993, §5.3; Dzombak and Hudson, 1995).

As pointed out by Stumm and Morgan (1996, §9.8), the "standardized" laboratory operations vary, and "it is difficult to come up with an operationally determined 'ion exchange capacity' that can readily be conceptualized unequivocally".

7.2.3   Isotherms

Ion-exchange data can also be represented in the form of isotherms, which are plots of the equilibrium concentrations of ions in the exchanger (solid) phase versus equilibrium concentrations of the same ions in the co-existing solution phase.

Distribution Coefficients and Retardation

Possibly the simplest way to describe the adsorption of an element is to define the ratio K_d as

$$K_d = \frac{S_i}{C_i} \tag{7.5}$$

where S_i is the mass of adsorbed i per unit mass of solid phase and C_i is the concentration of solute i in solution. It is, in a sense, a form of Henry's Law (§3.4.1). K_d has dimensions of L^3/M, commonly $ml\,g^{-1}$. If, in a groundwater system, a solute is thus partitioned onto solid phases, its movement is retarded with respect to the groundwater flow velocity. A *retardation factor* is defined as

$$R_i = 1 + \frac{\rho_b}{\theta} K_d \tag{7.6}$$

where ρ_b is the bulk density of the aquifer (M/L^3) and θ is the effective porosity (dimensionless). This factor is then easily incorporated into flow equations to describe the movement of adsorbed solutes. This procedure is so commonly used, and leads to so many errors, that we have given an expanded treatment of it in §10.3.

Isotherms

Relationships such as (7.5) which describe the distribution of solutes between adsorbing surfaces and the bulk solution are of course temperature dependent, and have come to be called *isotherms*. Equation (7.5) is a particularly simple example, called a *linear isotherm* in which S is linearly proportional to C, but there are several others.

The most obvious difficulty with the distribution coefficient (7.5) is that it predicts an unlimited capacity of the surface to adsorb solutes, whereas in reality there are a finite number of sites available, and, when these are occupied by solute ions, no further adsorption will occur. We should therefore expect that a plot of adsorbed vs. bulk solution concentrations should be curved, not linear. Assuming a fixed number of moles of identical adsorption sites (b), it can be derived that the moles of sites occupied by ion i, S_i, is

$$S_i = \frac{K_i b C_i}{1 + K_i C_i} \tag{7.7}$$

K_i is an ion-specific constant for reactions of the type

$$L + C_i = LC_i \tag{7.8}$$

where L is the concentration of a surface site arbitrarily named L, C_i is the concentration of an adsorbing ion (Ca^{2+} in the following examples), and LC_i is the concentration of sites L occupied by the adsorbing ion. In other words,

$$K_i = \frac{[LC_i]}{[L][C_i]} \tag{7.9}$$

where, as before, [ ] denotes concentration.[1] If 1 kg of water is understood, the terms can be in moles. b is thus the maximum value that LC_i can attain. A measure of site saturation is provided by θ, where

$$\theta = \frac{[LC_i]}{b} \tag{7.10}$$

which must vary from 0 to 1. Equation (7.7) is called the Langmuir isotherm. There is no difference between a Langmuir isotherm equilibrium constant and a surface complex formation constant, except that the Langmuir treatment does not explicitly involve electrostatic forces (see, e.g., Stumm and Morgan, 1996, §9.3).

Normally, isotherms are plotted as LC_i vs. C_i (see Figure 7.3a.), showing a curve asymptotic to b. Another way to plot it is *inverse LC_i* vs. *inverse C_i*. For pure Langmuir behavior, this will give a straight line with a slope of ($K_i^{-1} \cdot b^{-1}$) and an intercept of b^{-1} (see Figure 7.3b.) (Stumm and Morgan, 1996, §9.3).

Another commonly used relationship is the *Freundlich isotherm*,

$$S_i = K_F C_i^n \tag{7.11}$$

where K_F and n are constants. If $n = 1$, the above formula reduces to the linear adsorption isotherm; n is typically less than or equal to 1. It suffers from the same limitation of (7.5) in showing an unlimited capacity for adsorption, but the adjustable exponent means that it often fits experimental data quite well. It can be derived by considering surfaces to contain groups of sites, each group describable by a Langmuir isotherm.

It should also be noted that for small values of C (low aqueous concentrations), Equation (7.7) degenerates to Equation (7.5). That is, linear isotherms correspond to the low concentration region of the Langmuir isotherm.

---

[1] In computer programs, C_i is often an activity term, rather than a concentration. See §7.4.1.

Source	"Clean" surface	Typical surface complexes
Dzombak and Morel (1990)	$\equiv Fe^w OH^\circ$	$\equiv Fe^w OCu^+$
	$\equiv Fe^s OH^\circ$	$\equiv Fe^s OH_2^+$
Bethke (1996)	>(w)FeOH	>(w)FeOCu$^+$
(The Geochemist's Workbench™)	>(s)FeOH	>(s)FeOH$_2^+$
PHREEQC	Hfo_wOH	Hfo_wCu+
	Hfo_sOH	Hfo_sH2+

Table 7.1. Some ways of representing surface complexes on HFO. In programs such as The Geochemist's Workbench™ and PHREEQC, the user may define other types of notation and other surfaces. The default ones are shown.

7.2.4   Ion-exchange vs. Surface Complexation

Ion-exchange reactions and the CEC concept are used for the major ions in natural systems, rather than for minor and trace components. Although, for the most part, it is not known exactly where or how these major ions are held by the solid phases, it is probably best to envisage ion-exchange reactions as dealing with ions in the diffuse layer, rather than attached to the solid surface itself (Figure 7.1). Minor and trace element adsorption is now believed to be dominantly in the Stern layer, where they are attached to specific sites on the solid surface.

7.3   Surface Complexation

Just as exchange reactions make use of a fictive component X^-, so binding of elements to specific surface sites makes use of a new component for each type of site. The number of different types of sites on a mineral surface is not yet a measurable quantity, so it is, for the most part, at present, a parameter which varies between models, or between minerals within a model. For hydrous ferric oxide (HFO), for example, Dzombak and Morel (1990) concluded that two types of sites could accommodate all experimental data. These are the "strong" and "weak" sites, labelled "s" and "w" in Table 7.1.

A very useful point of view from a modeling standpoint is to consider that each type of site has a specific affinity, or a specific amount of attraction, for each type of ion in the solution, and that this can be quantitatively described by mass-action (equilibrium constant) equations. In other words, the solid sorption site behaves much like a complexing ligand in solution, and equations can be written involving sorption sites and ions, having $\Delta_r G$ values, activities and equilibrium constants. If we know the amount of surface area and the density of sites in that area, this gives the concentration ($\approx$ activity) of sorption sites. Then, if we have an equilibrium constant involving reactions between that site and various ions, we will have a system of equations that can be solved by our computer programs in the normal way to give the concentrations of surface species as well as those of the usual aqueous species.

Complexation Reactions

Consider a system consisting simply of a single solid phase in water. As discussed above, the surfaces of the solid phase will be electrically charged, even if there are no solutes in the water (Figure 7.1). However, the system as a whole is electrically neutral. This is because each charged surface induces a charge of an equal amount and an opposite sign in the adjacent solution. In other words, the solution adjacent to a negatively charged surface will have a preponderance of positively charged ions, and vice versa, the oppositely charged ones having being repelled by the surface. Whether the surface has a positive or negative charge depends largely on the solution composition and the nature of the ions adsorbed on the surface. In modeling, this charge can be calculated simply by adding up the charges on the various surface complexes.

Figure 7.2(a) shows a few of the possible arrangements. As mentioned above, the formation of each surface complex can be represented by a reaction. If the central metal ion (black circle) is Fe, the "clean" surface can be represented in various ways, depending on how we distinguish a surface complex from an aqueous complex. Some of these ways are shown in Table 7.1

The "clean" surface and the surface complex are related by reactions in the usual way, such as:

$$>(\text{w}) \, \text{FeOH}_2^+ = \, >(\text{w}) \, \text{FeOH} + \text{H}^+ \qquad \log K = -7.29 \qquad (7.12)$$

or

$$\equiv \text{Fe}^{\text{w}}\text{OH}_2^+ = \equiv \text{Fe}^{\text{w}}\text{OH}^\circ + \text{H}^+ \qquad \log K = -7.29 \qquad (7.13)$$

or, in PHREEQC,

$$\text{Hfo\_wOH} + \text{H}+ = \text{Hfo\_wOH2}+ \qquad \log K = 7.29 \qquad (7.14)$$

Note that PHREEQC writes equations in the opposite way to Dzombak and Morel (1990), a point to remember when altering databases (Appendix A).

But there is one additional factor in surface reactions, not present in aqueous phase reactions. Because the solid surface is electrically charged, it is an environment quite unlike the bulk solution, and activities and equilibrium constants must be adjusted to account for the fact that the surface sorption sites exist in a charged field.

7.3.1 The Electrical Double layer

As mentioned above, the existence of a charged surface layer induces an equal and opposite charge in the adjacent solution. This situation is described as the "electrical double layer" (EDL). The reactions controlling the charges on the surface may be modeled as described above, but the existence of the electrical field adds a complication not present in the usual aqueous reactions.

In such charged fields, chemical potentials, activities, and equilibrium constants are different from the values they would have in the absence of the charged field. By convention, equilibrium constants are given their values at the surface, before any

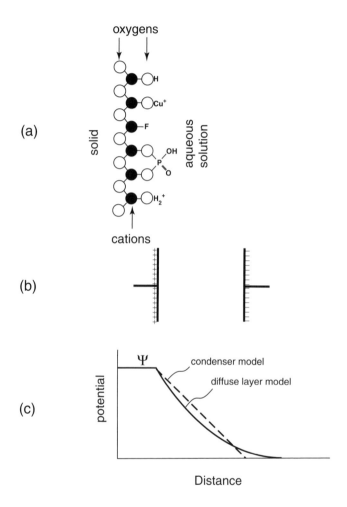

Figure 7.2. (a) Various types of surface complexes. A surface charge results from the combined positive and negative charged complexes. Adjacent to this surface in the aqueous solution, an equal and opposite charge is induced. (b) This doubly layered charge distribution can be modeled as a condenser. (c) More accurately, the potential resulting from the charged ions in the aqueous solution dies out gradually with distance from the surface, as in the diffuse layer model.

correction is applied (e.g., Equations (7.12) – (7.14)). The correction factor for the change in Gibbs energy of a mole of particles between the solution and the surface having a potential of Ψ volts is the energy (work) expended in transferring those particles from the aqueous solution to the surface. For a particle having a single electronic charge e, this work is $e \cdot \Psi$ (joules $=$ coulombs $\times$ volts). For a mole of such charges, it is $N_a e \cdot \Psi$, where N_a is Avogadro's number. $N_a e$ is the Faraday constant, $\mathcal{F}$, so the correction term for a mole of particles having a valence (number of electron charges) of z is $z \mathcal{F} \Psi$. In other words, for a mole of charged particles (ions),

$$G_{\text{solution}} = G_{\text{surface}} + z \mathcal{F} \Psi \qquad (7.15)$$

The correction factor for positive and negative ions will have opposite signs and will cancel. However, although individual surface complexation reactions (e.g., Equations (7.12) – (7.14)) are charge balanced, there is often a net change in charge of surface species (sites). For example, in Equation (7.12), $\Delta z = -1$. This is the number of charges that must be transferred from the solution to the surface, or vice versa. The $\Delta_r G^\circ$ of the surface complexation reaction will therefore have a correction factor of $\Delta z \mathcal{F} \Psi$, or

$$\Delta_r G^\circ_{\text{solution}} = \Delta_r G^\circ_{\text{surface}} + \Delta z \mathcal{F} \Psi \qquad (7.16)$$

Combining this with Equation (3.26),

$$\Delta_r G^\circ = -RT \ln K$$

in which $\Delta_r G^\circ$ and K can refer to $\Delta_r G^\circ_{\text{solution}}$ and K_{solution} *or* to $\Delta_r G^\circ_{\text{surface}}$ and K_{surface}, we have

$$-RT \ln K_{\text{solution}} = -RT \ln K_{\text{surface}} + \Delta z \mathcal{F} \Psi$$

so

$$\ln K_{\text{solution}} = \ln K_{\text{surface}} - \frac{\Delta z \mathcal{F} \Psi}{RT} \qquad (7.17)$$

or

$$K_{\text{solution}} = K_{\text{surface}} \cdot \exp\left(-\frac{\Delta z \mathcal{F} \Psi}{RT}\right) \qquad (7.18)$$

Modeling programs deal exclusively with equilibrium constants defined for the aqueous solution, so the "intrinsic" or "surface" equilibrium constants K_{surface} obtained from the literature (e.g., Dzombak and Morel, 1990) are corrected using Equations (7.17) or (7.18).

There remains one problem – what is the value of Ψ for a charged surface?

The Gouy–Chapman Model

Although the electrical double layer (EDL) can be modeled as a simple condenser (Figure 7.2b), the distance between the plates and the dielectric constant of the material between the plates must be known in order to relate the potential difference to the charge on the condenser. Neither of these quantities is easily determined in the present situation. The situation becomes even more difficult if the imaginary plate representing the aqueous solution is replaced by a more realistic gradual change in potential away from the surface (the diffuse layer, Figure 7.2c).

A solution to the diffuse layer problem was proposed independently by Gouy and by Chapman:

$$\sigma = (8RT\varepsilon\varepsilon_0 I \times 10^3)^{\frac{1}{2}} \sinh\left(\frac{z\Psi\mathcal{F}}{2RT}\right) \tag{7.19}$$

where σ is the surface charge density, ε is the dielectric constant of water, z is the ion charge, and ε_0 is the permittivity of free space. The original theory was developed for a single symmetrical electrolyte of concentration c. We follow Bethke (1996) in substituting the ionic strength I for c for the case of solutions with mixed electrolytes.

With known site densities, equilibrium constants for each site, and each solution component, and a way to relate surface potential to surface charge, the distribution of aqueous and surface species can now be solved in the same way as solutions with no surfaces. Bethke (1996) gives details of the method.

7.3.2 Other Surface Models

Mainly because of the work of Dzombak and Morel (1990) in providing a consistent theory and especially a self-consistent database of equilibrium constants, most widely available modeling programs use the double diffusive layer theory adopted by them. However, there are many other alternatives.

The simple condenser model or constant capacitance model has been mentioned above. It is essentially a simplified double layer theory, in which the distance between the plates can be calculated by using part of the Gouy–Chapman theory. See Stumm (1992) for details.

The triple layer model is, on the other hand, a slightly more complex version of the double layer theory, in which the surface layer is considered to be made up of two different layers – one closely bound to the surface, and one less closely bound. Several variations on this theme are to be found in the literature (Davis *et al.*, 1978).

7.4 Sorption Implementation in Computer Programs

Although there may be ambiguity in our understanding of "ion-exchange", and exactly what kinds of exchange are occurring in natural systems, there is no such ambiguity in the use of these terms in computer programs. Programs such as The Geochemist's Workbench™ and PHREEQC, which are the most versatile of presently available programs, implement these concepts in quite precise ways, which is not the same as saying that they are entirely correct, or that they will not be improved as our understanding of natural systems improves.

Programs RXN and REACT in The Geochemist's Workbench™ can account for the sorption of aqueous species onto mineral surfaces by several methods, including the two layer surface complexation model of Dzombak and Morel (1990), the constant

```
surface_data FeOH.dat
decouple Fe+++
swap Hematite for Fe+++
1 free gram Hematite
ph 5.3
2.4 mg/kg Ca++
3.5 mg/kg Mg++
0.04 mg/kg Ba++
0.4 mg/kg Na+
1.2 mg/kg Zn++
15.0 mg/kg SO4--
8 mg/kg Cl-
go
```

Table 7.2. A REACT script to calculate the amounts of various ions adsorbed by 1 g of
HFO. From The Geochemist's Workbench™ User's Guide.

capacitance and constant potential models, ion-exchange, distribution coefficients (K_d),
and Langmuir isotherms. Program PHREEQC also includes the Dzombak and Morel
formulation and data, and ion-exchange reactions (and data). Distribution coefficients
and Langmuir isotherms are not explicitly included, but can be simulated nonetheless.

7.4.1   Examples

To show how these programs work, we consider a couple of examples from The Geochemist's Workbench™ User's Guide, then repeat them with PHREEQC. Examples of
more complex systems are considered in later chapters.

The Dzombak and Morel Model for Adsorption on HFO

The Geochemist's Workbench™   Consider a dilute solution containing several ions in
contact with the adsorbing surfaces of hydrous ferric oxide (HFO). A script for program
REACT is shown in Table 7.2.
     The Geochemist's Workbench™ stores the Dzombak and Morel data in a file called
FeOH.dat. The program calculates the sorbing surface area and available sites from
the mass of mineral (1 g) and data in FeOH.dat.

PHREEQC   A script to perform the same calculation in PHREEQC is shown in Table 7.3.
The Dzombak and Morel data are included in the default PHREEQC database. In Table 7.3,
the key word SURFACE defines the surface properties to be the same as those in the
REACT example, and EQUILIBRIUM_PHASES declares that hematite is in equilibrium

```
TITLE Example from GWB User's Guide
SURFACE 1
        -equilibrate with solution 1
        Hfo_sOH         6.262e-5    600.    1.0
        Hfo_wOH         2.505e-3
EQUILIBRIUM_PHASES 1
    Hematite    0.  0.006262
SOLUTION 1
    -units mg/kgw
    ph      5.3
    Ca      2.4
    Mg      3.5
    Ba      .04
    Na      0.4
    Zn      1.2
    S(6)        15.
    Cl      8       charge
SELECTED_OUTPUT
    -file   sel_HFO.out
    -reset false
USER_PUNCH
    -headings  Ba  Ca SO4   Zn
    -start
    10 sorbed_Ba = mol("Hfo_sOHBa+2")   + mol("Hfo_wOBa+")
    20 sorbed_Ca = mol("Hfo_sOHCa+2")   + mol("Hfo_wOCa+")
    30 sorbed_SO4= mol("Hfo_wOHSO4-2")  + mol("Hfo_wSO4-")
    40 sorbed_Zn = mol("Hfo_sOZn+")     + mol("Hfo_wOZn+")
    50 frac_Ba   = sorbed_Ba/ (sorbed_Ba + TOT("Ba"))
    60 frac_Ca   = sorbed_Ca/ (sorbed_Ca + TOT("Ca"))
    70 frac_SO4  = sorbed_SO4/(sorbed_SO4 + TOT("S(6)"))
    80 frac_Zn   = sorbed_Zn/ (sorbed_Zn + TOT("Zn"))
    90 PUNCH frac_Ba  frac_Ca  frac_SO4  frac_Zn
    -end
END
```

Table 7.3. A PHREEQC script to calculate the amounts of various ions adsorbed by 1 g of HFO.

	Fraction sorbed	
	REACT	PHREEQC
Ba	0.003028	0.003036
Ca	0.0009716	0.0009721
SO4	0.8477	0.8477
Zn	0.4018	0.4022

Table 7.4. Comparison of sorption results from REACT and PHREEQC.

with the solution (SI = 0.0). The USER_PUNCH keyword is useful here, because we want results in the same form that REACT provides, which is the fraction of element sorbed, where the total amount of the element is the amount sorbed *plus* the original solution molality. Note that, unless instructed otherwise, both programs calculate the amount sorbed in equilibrium with the solution composition given, so the total amount of element after sorption can be quite a bit greater than in the original solution. USER_PUNCH allows us to print out a file with this information.

The results from the two programs are shown in Table 7.4.

The Langmuir Isotherm

The Geochemist's Workbench™   Program REACT calculates the amount of elements adsorbed according to the Langmuir isotherm by first reading a user prepared file which gives the adsorption reactions and their equilibrium constants. The format of this file is given in The Geochemist's Workbench™ distribution in an example file to use as a template. In this file, both Ca and Sr are given reactions and constants for a single adsorption site called L. In our example, we have used this file, but eliminated the Sr adsorption, so that only Ca adsorbs on the site L, according to the reaction

$$L + Ca = LCa \qquad \log K = 2.1 \qquad (7.20)$$

where L is the concentration of surface sites, Ca is the *activity* of the calcium ion, and LCa is the adsorbed concentration of calcium.

The exchange capacity (the total number of sites, and therefore the maximum amount of adsorbed Ca) in moles (0.01) is given in the input script. The script shown in Table 7.5 is taken from the User's Guide, substituting our new file name for the modified surface data. Because we want to use a number of different Ca concentrations, we must modify the script and run it repeatedly.

PHREEQC   Program PHREEQC does not contain an explicit provision for calculating Langmuir isotherms, but has enough options in the SURFACE data block to do the same thing. A script to perform the same calculation done by REACT is shown in Table 7.6.

In this script, the data blocks SURFACE_MASTER_SPECIES and SURFACE _SPECIES perform the same function as the separate file we prepared for REACT, langmuir2.dat. They define a surface site L, and give the equilibrium constant for the adsorption reaction. The SURFACE data block equilibrates the solution

```
surface_data = Langmuir2.dat
pH = 8.3
Cl-   = 19000 mg/kg
Ca++  =   400 mg/kg
Sr++  =    10 mg/kg
Mg++  =  1300 mg/kg
Na+   = 10700 mg/kg
K+    =   400 mg/kg
SO4-- =  2700 mg/kg
exchange_cap = 0.01 moles
go
```

Table 7.5. A REACT script to calculate adsorption of Ca according to the Langmuir isotherm.

```
TITLE Langmuir isotherm (from GWB) SURFACE_MASTER_SPECIES
        L          L
SURFACE_SPECIES
        L = L
          log_k     0.0
        L + Ca+2 = LCa+2
          log_k     2.1
SURFACE 1
        -equilibrate with solution 1
        -no_edl
        L    0.01
SOLUTION 1
        -units mg/kgw
        ph        8.3
        Ca        10000
        Mg        1300
        Na        10700
        Sr        10
        K         400
        S(6)      2700
        Cl        19000          charge
END
```

Table 7.6. A PHREEQC script to calculate Ca adsorption according to a Langmuir isotherm.

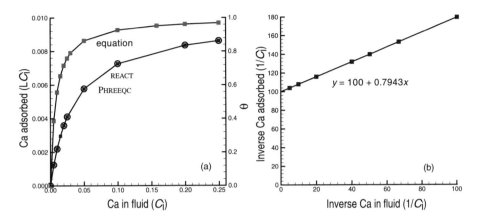

Figure 7.3. (a) Comparison of the Langmuir isotherm calculated by REACT, PHREEQC, and Equation (7.7). (b) Inverse plot and least squares fit for the equation results.

(SOLUTION 1) with this surface and contains the instruction -no_edl, which turns off all electrostatic calculations, those necessary for the electrostatic double layer theory. It also defines the total number of L sites. There is no explicit mention of Langmuir isotherm because, as mentioned above, there is no difference between a Langmuir isotherm equilibrium constant and a single-site surface complex formation constant, which PHREEQC understands.

This script must be run several times, varying the Ca content to define the complete isotherm. Alternatively, the script can be extended, with each calculation separated by an END statement. Using the script shown in Table 7.6, the program prints a voluminous output file which contains a lot of data besides the adsorption data. We could add the following statements to make the task of finding these data easier.

```
SELECTED_OUTPUT
        -file    lang.sel
        -reset   false
USER_PUNCH
        -headings  Total_Ca  Ca+2  Adsorbed_Ca
        -start
        10 PUNCH TOT("Ca")   MOL("Ca+2") MOL("LCa+2")
        -end
```

These cause another output file to be written, lang.sel, containing only the total Ca, the Ca^{2+} molality, and the moles of adsorbed Ca.

The calculated isotherms from REACT and from PHREEQC are virtually identical, as shown in Figure 7.3(a). Note that although the two programs give the same results, these are quite different from those obtained from Equation (7.7) with the same data $(K_i = 10^{2.1}; b = 0.01; C_i = 0 \rightarrow 0.25)$, although all results tend towards a maximum of 0.01, or $\theta = 1$, as they should. This is because both programs use the *activity* of Ca^{2+} in their calculations, rather than the nominal concentration. The activity coefficient of Ca^{2+} is about 0.2 under these conditions, which accounts for the results from the two programs being a factor of about 0.2 less than the results from the equation (see §3.4.1 for the activity–concentration relationship).

The expected results for the inverse plot (Figure 7.3b) are

$$\text{intercept} = b^{-1} = 1/0.01 = 100, \text{ and}$$

$$\text{slope} = K^{-1} \cdot b^{-1} = (1/10^{2.1}) \cdot 100 = 0.7943.$$

The inverse plot gives the expected results for the equation results. An inverse plot of the computer program results gives a straight line, but with a different slope and intercept.

7.4.2   Why Surface Modeling is Not Perfect

Dzombak and Morel (1990) performed a great service for surface modelers in systematizing a huge body of diverse literature on hydrous ferric oxide (HFO) research, and presenting it in a form which readily lent itself to use by programmers. In doing so, they made a number of decisions about HFO, such as the surface area, number of active sites, site densities, and so on (Table 7.7), as well as of the type of model and the data to use. All these choices find their way into the default values of these parameters of the programs written subsequently. There is nothing intrinsically wrong with these choices; they may well be the best possible overall values. However, they cannot be the best values for all possible applications.

For one thing, hydrous ferric oxide, although a very important adsorber in natural systems, is an extremely complex substance, having a wide range of physical properties (see Cornell and Schwertmann (1996) for a recent summary). Therefore, it is never certain that the default properties of HFO or the experimental data on HFO chosen by Dzombak and Morel (1990) will be appropriate for given natural systems. Also, the amount of this phase present is often poorly known.

It must also be remembered that HFO is only one of the many phases in natural samples, and although an important adsorber, it is not the only one. However, it is the only one in most models. Possibly, other phases, although weak adsorbers, could play important roles by being present in large amounts.

Furthermore, most experimental work applies to simple laboratory systems, and a great deal more work will be needed to understand adsorption quantitatively in complex natural systems. Particularly lacking is a mechanistic understanding of the role of the solid phase, and any ability to predict surface complexation constants. A promising start in this direction is provided by Sverjensky (1993, 1994), and Sverjensky and Sahai (1996).

7.5   Retardation of Radionuclides at Oak Ridge

Saunders and Toran (1995) used the diffusive double layer model of Dzombak and Morel (1990) to model the role of HFO in retarding radionuclide migration at the Oak Ridge Reservation, Tennessee. Large amounts of low-level radioactive wastes were disposed into shallow, unlined lagoons and trenches, resulting in concerns for their potential migration. Saunders and Toran (1995) used MINTEQA2 to perform speciation, solubility,

Site density:		
strong	$5.62 \times 10^{-5} \, \text{mol g}^{-1}$	$5 \times 10^{-3} \, \text{mol mol}^{-1}$
weak	$2.25 \times 10^{-3} \, \text{mol g}^{-1}$	$0.2 \, \text{mol mol}^{-1}$
Surface area	$600 \, \text{m}^2 \, \text{g}^{-1}$	$600 \, \text{m}^2 \, \text{g}^{-1}$
Molecular formula weight	$89 \, \text{g mol}^{-1}$	$89 \, \text{g mol}^{-1}$

Table 7.7. Surface properties of hydrous ferric iron oxides. Data from Dzombak and Morel (1990).

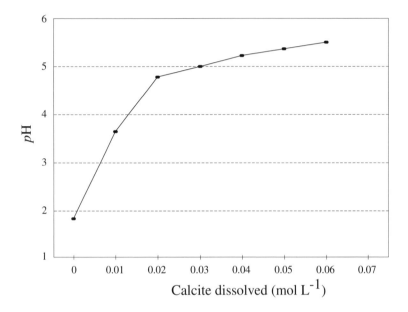

Figure 7.4. Results of MINTEQA2 calcultions in Saunders and Toran (1995).

and surface complexation calculations and compared the modeling results with field data. In one of their modeling applications, the acidic water from S-3 pond, Y-122 was allowed to dissolve calcite – a reaction assumed to be the dominant neutralization mechanism in the aquifer. Aqueous speciation, Saturation Indices of minerals, and surface complexation were calculated at each pH increment. Apparently, supersaturated solids were not allowed to precipitate in their model runs. An average amount of amorphous HFO of 0.9 wt.% was used for calculating the sorbent concentrations, and the surface properties recommended by Dzombak and Morel (1990) were used for their amorphous iron oxide (Table 7.7). The intrinsic complexation constants of Dzombak and Morel (1990) were added to the MINTEQA2 surface complexation reaction database.

The modeling results are shown in Figures 7.4 and 7.5. MINTEQA2 calculations show possible solubility control of Fe and Cd by precipitation of siderite ($FeCO_3$) and otavite ($CdCO_3$), and significant retardation of heavy and transition metals by HFO in the order of Pb > Zn > Ni > ^{60}Co. The model also predicts little sorption of ^{90}Sr by Fe oxyhydroxide at $pH < 7$. The modeling results illustrate the pH-dependent nature

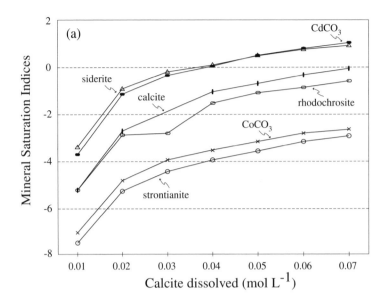

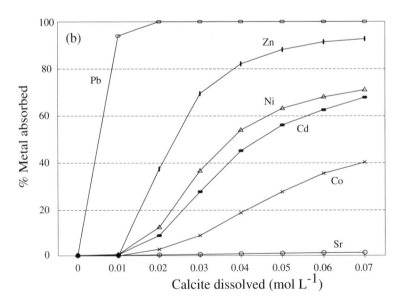

Figure 7.5. Saturation Indices (a) and surface adsorption (b) calculated from MINTEQA2 in Saunders and Toran (1995).

Calculation of Sorbent Concentrations: Example 1

Saunders and Toran (1995) used an average of 0.9 wt.% amorphous $Fe(OH)_3$ in soil around the Oak Ridge burial ground for calculating sorbent concentrations. The amorphous iron concentrations were obtained from the sequential extraction techniques. To calculate the surface sites in their surface complexation modeling, they used a soil porosity of 40% and dry density of $2.5\,g\,cm^{-3}$, respectively. The calculation is recapped here:

For a volume of $1000\,cm^3$ aquifer, the solid matrix portion has a volume of

$$1000\,cm^3 \times (1 - 40\%) = 600\,cm^3$$

The weight is

$$600\,cm^3 \times 2.5\,g\,cm^{-3} = 1500\,g$$

The total $Fe(OH)_3$ in the soil matrix is

$$1500\,g \times 0.9\% = 13.5\,g$$

The volume of pore water in a $1000\,cm^3$ aquifer is $400\,cm^3$. The amount of $Fe(OH)_3$ in contact with 1.0 liter of water is

$$13.5\,g/0.40 = 33.75\,g\,L^{-1}\ as\ Fe(OH)_3$$

With a formula weight of 89 g,

$$the\ sorbent\ concentration = 33.75/89 = 0.34\,mol\,L^{-1}$$

of metal sorption onto HFO. Comparison with field monitoring data shows that the model-predicted relative mobilities are basically consistent with field data. It should be noted that a number of assumptions are made for the modeling runs. For example, Saunders and Toran (1995) discussed the omission of the possible sorption of metals to precipitating Fe and Al oxyhydroxides.

7.6   Mobility of Radionuclides at a Uranium Mill Tailings Impoundment

In the discussion of the Bear Creek Uranium site in §6.2, we note that more than a dozen regulated metals and radionuclides have elevated concentrations in the acidic plume. The relative mobility of the radionuclides at the leading edge of the plume can be estimated by using the diffusive double layer model of Dzombak and Morel (1990).

Calculation of Sorbent Concentrations: Example 2

Parkhurst (1995, example 10) also did calculations of sorbent concentrations in his surface complexation modeling for the central Oklahoma aquifer. The amount of extractable iron in sediments is from 1.6 to 4.4%. Porosity is 22%, and rock density is $2.7 \, g \, cm^{-3}$. For a 2 wt.% iron, the solid concentrations should be calculated as follows. The total weight for a volume of $1000 \, cm^3$ aquifer is

$$1000 \, cm^3 \times (1 - 22\%) \times 2.7 \, g \, cm^{-3} = 2106 \, g$$

The total iron in the solid is

$$2106 \, g \times 2\% = 42.12 \, g$$

The amount of iron in contact with 1.0 liter of pore water is

$$\text{solid concentration} = 42.12 \, g/0.22L = 192 \, g \, L^{-1} \text{ as Fe}$$

or

$$\text{solid concentration} = 306 \, g \, L^{-1} = 3.4 \, mol \, L^{-1} \text{ as Fe(OH)}_3$$

Parkhurst (1995) assumed 10% extractable iron is in amorphous form so that the sorbent concentrations he used is $0.38 \, mol \, L^{-1}$.

7.6.1 Why Geochemical Modeling?

At the Bear Creek Uranium site, groundwater monitoring data have been collected on a quarterly basis. These data outline the distribution of the hazardous constituents over the time and provide an observational basis on the mobility of the radionuclides under the site's geologic conditions. Geochemical modeling can provide interpretations of the observational data so that prediction of potential migration in the future can be built on a scientific basis. For hazardous constituents that show no or limited mobility at the site, geochemical modeling can provide a plausible interpretation, and, hence, strengthen the case for eliminating them in further considerations. For hazardous constituents that show migration at the site, quantitative predictions of their concentrations in the future have to be provided to the decision makers.

7.6.2 Modeling Approach

As discussed in §6.2, the migrating acid plume has a pH of about 4.5, buffered by the precipitation of amorphous Al(OH)$_3$, and carries with it the elevated concentrations of radionuclides and hazardous constituents. When the acidic water encounters calcite in the aquifer in the downgradient area, calcite dissolution occurs, and that buffers the

Al	3.79×10^{-7}	Mg	4.19×10^{-2}
Ba	1.48×10^{-7}	Mn	1.27×10^{-3}
C	5.0×10^{-2}	Na	1.27×10^{-2}
Ca	6.73×10^{-3}	SO4	3.27×10^{-2}
Cl	1.15×10^{-2}	Si	3.62×10^{-4}
F	1.07×10^{-5}	Gypsum	0.082
Fe	1.6×10^{-8}	Al(OH)$_3$	0.013
K	1.82×10^{-3}	Fe(OH)$_3$	0.038
pH	6.6		

Table 7.8. Composition of MW-77 in $mol\,L^{-1}$ after being neutralized by 0.08 mole calcite.

groundwater pH to about 6.5.

Therefore, we take two steps in modeling. First, the acidic water is neutralized by titrating of calcite. The amounts of Fe and Al hydroxides precipitated from this titration are then used in the subsequent speciation and complexation modeling.

In the first step, acidic water of about pH 4.5 was first neutralized to a pH of 6.6 by titrating calcite using PHREEQC (see §8.3). Neutralization of the acid water caused the precipitation of gypsum, and Al and Fe hydroxides. This kind of reaction path modeling will be discussed in Chapter 8. Here, we only present the resulting compositions of the neutralized water and precipitated solids in Table 7.8.

In the second step of the modeling exercise, speciation and surface adsorption in the neutralized water were modeled using the geochemical code MINTEQA2. Analytical concentrations of Cd, Ni, Be, and U (5.5, 65, 4.1, and $22 \times 10^{-6}\,mol\,L^{-1}$, respectively) were used as input concentrations. The surface properties and surface complexation constants were taken from Dzombak and Morel (1990). Due to the lack of experimental data for Al hydroxides, we assume that all Al hydroxides have the same sorptive properties as Fe hydroxide and that the mass of Al hydroxide was added to the Fe hydroxide concentrations. The total sorbent concentration is 5.3 g/L as Fe(OH)$_3$, which is calculated from the first step. MINTEQA2 was used to calculate the partitioning of these ions between the aqueous phase and ferric iron hydrous oxide (HFO) surfaces.

7.6.3   Modeling Results

Tables 7.9 and 7.10 show the results from MINTEQA2. The model predicts that the dominant aqueous species for uranium and nickel in water are carbonate complexes. A significant proportion of sorption sites on the ferric oxides are occupied by sulfate ions, which results in fewer sorption sites available for metals and radionuclides.

7.6.4   Comparison with Field Data

Comparing the modeling results with site monitoring data shows that the predicted sequence of the contaminant mobility, Be < Cd < Ni < U, is consistent with the metal distribution at the site. Surface complexation has essentially immobilized Be and Cd

Component	Species	Distribution (%)[a]
Be^{2+}	Be^{2+}	100 [b]
Ni^{2+}	$NiCO_3(aq)$	88
	$NiHCO_3^+$	12
Cd^{2+}	Cd^{2+}	100
UO_2^{2+}	$UO_2(CO_3)_2^{-2}$	76
	$UO_2(CO_3)_3^{-4}$	24
Sorption site 1	$\equiv SO1HCa^{2+}$	45
	$\equiv SO1HSO_4^{-2}$	17.3
	$\equiv SOH1^\circ$	13
	$\equiv SO1H2^+$	11
	$\equiv S1SO_4^-$	7
	$\equiv SO1Ni^+$	3
	$\equiv SO1Be^+$	2
Sorption site 2	$\equiv SO2HSO_4^{-2}$	34
	$\equiv SOH_2^\circ$	26
	$\equiv SO2H_2{}^+$	22
	$\equiv S2SO_4^-$	15
	$\equiv SO2UO_2^+$	2

Table 7.9. Species distribution in the aqueous phase and on sorption sites, as calculated by MINTEQA2.

[a]Aqueous speciation is normalized to aqueous phase only.
[b]No other aqueous Be species available in the database.

(100% and 98% sorbed onto the surfaces, respectively). Surface adsorption onto HFO that amounts to precipitation from one pore volume of fluid can control Be and Cd to <0.01 mg L^{-1}, the groundwater protection standard. The observational data show that Be is not seen outside of the acid plume, and Cd has limited migration outside of the plume. Modeling results show that 87% Ni is sorbed onto Fe oxyhydroxide at the time of neutralization, but significant concentrations remain in the aqueous phase. This is consistent with the monitoring data that show significant transport of nickel. The modeled equilibrium concentration of 4.8 mg L^{-1} is close to the measured concentrations at the edge of the current acid plume.

However, the predicted uranium concentration at the edge of the plume is far higher than the observation. Observational data indicate that the maximum uranium concentration outside of the plume is less than 92 pCi/L, while the model predicts it to be in excess of 1000 pCi/L. This discrepancy between observed and predicted uranium concentrations is an indication that the model parameters used to predict U partitioning may not accurately represent conditions at the site. Examination of Dzombak and Morel's (1990) work reveals that they have used a predicted surface complexation constant rather than retrieve a value from experiments. Newer experimental data show that uranium(VI) adsorption onto ferrihydrite can be fitted by a two-site model with a bidentate complex (Waite et al., 1994). However, this type of modeling ability is not included with MINTEQA2. Another explanation could be that co-precipitation is ignored, but apparently is a major attenuation mechanism at a similar site (Opitz et al. 1983).

Species	Dissolved (mol kg^{-1})	(%)	Sorbed (mol kg^{-1})	(%)
UO_2^{2+}	2.067×10^{-5}	94	1.325×10^{-6}	6
Al^{3+}	3.800×10^{-7}	100	0.000×10^{-1}	0
Ba^{2+}	8.083×10^{-8}	53.9	6.917×10^{-8}	46.1
CO_3^{2-}	5.000×10^{-2}	100	0.000×10^{-1}	0
Ca^{2+}	5.671×10^{-3}	84.6	1.029×10^{-3}	15.4
Cl^-	1.140×10^{-2}	100	0.000×10^{-1}	0
F^-	1.00×10^{-5}	100	0.000×10^{-1}	0
Fe^{3+}	1.600×10^{-8}	100	0.000×10^{-1}	0
K^+	1.820×10^{-3}	100	0.000×10^{-1}	0
Mg^{2+}	4.180×10^{-2}	100	0.000×10^{-1}	0
Mn^{2+}	1.270×10^{-3}	100	0.000×10^{-1}	0
Na^+	1.260×10^{-2}	100	0.000×10^{-1}	0
PO_4^{3-}	2.316×10^{-10}	0.2	9.977×10^{-8}	99.8
SO_4^{2-}	3.212×10^{-2}	98.2	5.838×10^{-4}	1.8
H_4SiO_4	3.620×10^{-4}	100	0.000×10^{-1}	0
Be^{2+}	7.703×10^{-9}	0	4.099×10^{-5}	100
Cd^{2+}	1.094×10^{-7}	2	5.391×10^{-6}	98
Ni^{2+}	8.240×10^{-6}	12.7	5.676×10^{-5}	87.3
H_2O	1.318×10^{-6}	-0.3	-4.862×10^{-4}	100.3
E^-	0.000×10^{-1}	0	0.000×10^{-1}	0
H^+	6.327×10^{-2}	99.5	3.308×10^{-4}	0.5

Table 7.10. Results of MINTEQA2 calculations of adsorption onto hydrous ferric iron oxide (HFO).

7.6.5   Discussion of Modeling Results

Are the modeling results practically useful? In this case, they are. Surface complexation and speciation modeling is shown to be an effective screening tool for evaluating the mobility of metals and radionuclides at the site. This screening resulted in an explanation for the mechanism of natural attenuation for hazardous constituents that are judged to be immobile from observation data. Elimination of these constituents from further analysis, e.g., solute transport modeling, can result in cost savings.

However, as we noted early in this chapter, numerous assumptions are employed in the field applications of surface complexation models. Davis *et al.* (1998) noted that surface complexation models are mainly developed from well-controlled laboratory experiments. It is unclear how the models can be applied to soil and sediments where the double layers of the heterogenous particles may interact and the competitive adsorption of many different ions can cause significant changes in the electrical properties of mineral–water interfaces.

7.7   Adsorption of Arsenic in Smelter Flue Dust

Doyle *et al.* (1994) used the triple layer model of Davis *et al.* (1978) to model surface adsorption of arsenic onto amorphous ferric oxides. Copper smelting has produced

Surface area	$200 \ \mathrm{m^2 \, g^{-1}}$
Site density	$11.4 \ \mathrm{site \, nm^{-2}}$
Inner layer capacitance	$1.4 \ \mathrm{F \, m^{-2}}$
Outer layer capacitance	$0.2 \ \mathrm{F \, m^{-2}}$

Table 7.11. Surface properties for the triple layer model for amorphous Fe oxide from Davis *et al.* (1978).

metal bearing flue dust as a by-product of smelter exhaust streams. This flue dust contains high concentrations of hazardous metals such as arsenic (16 500–79 200 ppm), zinc (19 200–49 000 ppm), lead (9 260–32 300 ppm), copper (69 400–244 000 ppm), etc. Safe disposal of these wastes into the environment is a concern. To evaluate disposal practices, Doyle *et al.* (1994) conducted column experiments. Flue dust waste cake from the Cashman hydrometallurgical processes was packed into glass columns and the columns leached with de-ionized water. Two types of flue dust were used to fill the columns: one from a batch reactor and one from a continuous reactor. The waste cakes were analyzed for their mineralogical compositions using an electron microprobe, and leachate was collected for analysis.

The authors first used MINTEQA2 to evaluate the solubility control of As concentrations in the effluents of column experiments. They found that it was necessary to replace the $\log K$ values for the arsenic aqueous species and the mineral scorodite [$FeAsO_4 \cdot 2H_2O$] in the MINTEQA2 database with those for the Fe–As–SO$_4$ systems from Robins (1990) in order to match the simulation and experimental results. From the microprobe study they found 1 μm-sized, As-bearing phases having a stoichiometric composition close to scorodite.

Sorption of $HAsO_4^{2-}$, $H_2AsO_4^-$ and SO_4^{2-} was simulated using MINTEQA2. The sorption of cations was not simulated. Surface properties and surface complexation constants were taken from Davis and Leckie (1978, 1980) and Davis *et al.* (1978) (Table 7.11). Varying the surface parameters over the ranges given in the original papers did not substantially change the modeling results. The sorbent concentration is assumed to be 1% of the total Fe in the flue dust present as $Fe(OH)_3$. The authors claim that variation of the sorbent concentrations up to two orders of magnitudes did not change the simulated results (mostly solubility control, not surface adsorption; see below).

The authors found that the model-predicted As concentration is close to the leachate concentrations from the column packed with dust from the continuous reactor (1 120 μg L^{-1} versus 1 330 μg L^{-1}), when the solubility product of scorodite from Robins (1990) and the triple layer model is used. Only 11% As in the system was sorbed onto hydrous ferric oxide surfaces. Arsenic concentrations in the leachate are largely controlled by scorodite solubility. It should also be pointed out that simulations using solubility only and without including surface adsorption resulted in a closer match (1 270 μg L^{-1} versus 1 330 μg L^{-1}). For the simulation of the column experiments using wastes from the batch-reactor, the triple layer model predicted too low an As concentration (33 μg L^{-1} versus 120 μg L^{-1}).

8

Reaction Path Modeling

8.1   Introduction

What is today called reaction path modeling was introduced to geochemistry by Helgeson (1968). It uses the concept of the *progress variable*,[1] proposed by deDonder in the 1920s and used extensively by Prigogine and Defay (1965), but, before Helgeson, almost unknown in geochemistry. The mathematical foundations are not difficult, but we omit this aspect here, and try instead to convey a sense of what the modeling achieves and does not achieve. Readers are referred to Anderson and Crerar (1993, §19.3), Helgeson (1979), Helgeson *et al.* (1970), and Wolery (1992) for discussions of the theory and mathematical formulations of the models. It was introduced in general terms in §2.3.2.

The general aim is to be able to trace what happens during *irreversible* reactions or processes, such as dissolution or precipitation of minerals, mixing of solutions, or cooling or heating of complex systems. If the reader has absorbed Chapter 3, or any similar material, one will suspect that this cannot be done with thermodynamics, because it can only deal with equilibrium states. Irreversible reactions might *begin* in a metastable equilibrium state (say a crystal of K-feldspar and a liter of pure water, separated) and *end* in a stable equilibrium state (the K-feldspar, plus alteration products, such as kaolinite, at equilibrium in a solution containing K, Al, and Si), but in between these states the system must necessarily pass through states of disequilibrium. How can thermodynamics deal with this?

It can do this not by trying to consider states of disequilibrium, but by being applied to a succession of states of metastable equilibrium. So in the K-feldspar case, a tiny amount of $KAlSi_3O_8$ is conceptually added to pure water, and the resulting aqueous species are calculated.[2] We have a new metastable state: K-feldspar and a very dilute solution of $KAlSi_3O_8$ in water.[3] Then another tiny increment of $KAlSi_3O_8$ is added to

---

[1]See Chapter 11 for another application of the progress variable.

[2]The increments don't always have to be tiny. They do in this case because K-feldspar is very insoluble, and many phase changes are involved.

[3]We call this a *metastable state* in the sense used in §3.1.1 and §3.1.2, that is, a state in a conceptual or model system, where the K-feldspar is separated from, or constrained from reacting with, the K-feldspar

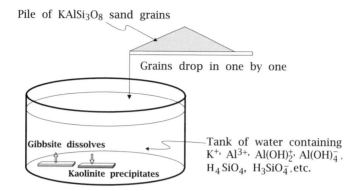

Figure 8.1. A possible conceptualization of the reaction path model for the dissolution of K-feldspar in water. (From Anderson, 1996, Chap. 12.)

the solution, and the new species concentrations and activities are calculated, and the solution *progresses* towards equilibrium with K-feldspar. After each increment is added, the Saturation Index (SI) of all minerals in the database containing any combination of K, Al, Si, and O is calculated. If any are superaturated, their composition is subtracted from the solution composition ("precipitated") until SI = 0 (equilibrium). Later, in the succession of additions of $KAlSi_3O_8$, some minerals previously precipitated may redissolve. What is done is entirely conceptual and mathematical, but one way to think about it is illustrated in Figure 8.1. Thus, thermodynamics is called upon to calculate the equilibrium state of a succession of solutions, until one is found in which equilibrium with K-feldspar is achieved. No kinetics or time scale is involved.

In this succession of equilibrium states, a surprising amount of precipitation and dissolution of phases is found to be involved, the details of which can be found in many geochemistry textbooks. But the question arises – is this succession of calculated states what would be observed if K-feldspar was *really* dissolved into water? Well, probably not. So why do the calculation? This is a crucial point which must be understood.

A wide variety of irreversible processes can be treated in this way. It is basically a *titration* of one composition into another composition. Quite often it is thought of as a very *slow* titration, giving the system enough time to adjust to each new equilibrium state. But what the results simulate is one irreversible path out of what is theoretically an infinite number of possible irreversible paths. Sometimes, as in the alkalinity example (§8.2), the results are probably very close to what actually happens and, therefore, are directly useful. In other cases, especially those involving minerals and fluids at ambient temperatures, as in the K-feldspar example, what actually happens might be different. Nevertheless, the calculation is always useful in understanding the system and the changes which are possible in it. The K-feldspar case actually sparked several years of research on what actually happens on feldspar surfaces in the weathering environment, and it still forms a framework for thinking about this system today, although no one takes the titration results literally.

(Figure 8.1). It is often referred to as a state of *partial equilibrium*.

```
TITLE Titrate MW-36 with HCl to get alkalinity SOLUTION 1 MW-36
   pH 7.4
   unit mg/l
   temp 15.0
   Na 61.
   K 17.
   Ca 158.
   Mg 21.
   Cl 25.
   Alkalinity 153.
   S(6) 425. charge
   Fe(3) 0.1
   Al 0.01
   Mn(2) 0.11
   Si 5.6
REACTION 1 Add HCl to the sol'n
   HCl 1.0
   0.004 moles in 20 steps
SELECTED_OUTPUT
   -file mw36_hcl.out
END
```

Table 8.1. PHREEQC input file for titrating MW-36 with HCl.

8.2 Alkalinity Titration

In this section we illustrate the computed alkalinity titration discussed in Chapter 3. We use PHREEQC to perform the calculations, using sample MW-36 from the Bear Creek Uranium site (see Chapter 6). The idea is

1. to use the alkalinity reported in the analysis to determine the total carbonate content of the solution, $m_{CO_3,total}$, and

2. to illustrate the acidity concept.

Alkalinity Titration of MW-36

The PHREEQC input file is shown in Table 8.1. Note that we use sulfate for charge balance, because it is the most abundant ion. The default definition of alkalinity in PHREEQC is "as calcite", which is what our analysis shows, so we simply enter that number ($153 \, mg \, L^{-1}$). Under the keyword REACTION, a suitable combination of moles of HCl and number of steps is found with some calculation and a certain amount of trial and error. We specify the file name of the output file to ensure we get an output file with the essential data arranged for copying to a spreadsheet. The results are shown in Figure 8.2.

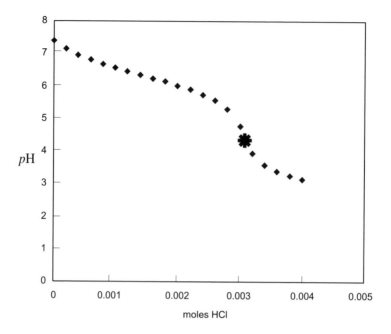

Figure 8.2. Results of the alkalinity titration of MW-36 with PHREEQC. The asterisk symbol represents the end-point of the titration, from which the alkalinity can be calculated.

The end-point at pH 4.3 occurs at about 0.0031 moles HCl, giving an alkalinity of (following the example calculation on p. 63) $0.0031 \times 100.08 \times 0.5 = 0.155 \, \text{g L}^{-1}$, or 155 mg L^{-1}. This is almost exactly the analyzed alkalinity of 153 mg L^{-1}, meaning that carbonate anions dominate in this solution. However, this does not give us the total carbonate content.

Using Equations (3.68)–(3.70) with $A = 0.00153$, $pH = 7.4$, the results are

$$m_{H_2CO3°} = 0.00023$$
$$m_{HCO_3^-} = 0.00305$$
$$m_{CO_3^{2-}} = 7.16 \times 10^{-6}$$

and

$$m_{CO_3, total} = 0.00328$$

As expected, with a pH of 7.4, $m_{HCO_3^-}$ dominates.

We don't present the results here, but in fact if we titrate the same solution (MW-36) with NaOH to pH 8.3, we find that about 2.5×10^{-4} moles NaOH are required, almost exactly the $m_{H_2CO_3°}$ content found by the calculation, as it should be. This confirms the point made in Chapter 3 that the sum of the alkalinity and the acidity titration is equal to the total carbonate content (in the absence of competing acids and bases, and assuming $pH \approx 4.3$ and ≈ 8.3 end-points).

8.3   Acidity of Acid Mine Water

Here, we use the Bear Creek Uranium example discussed in §6.2 to calculate the acidity of contaminated groundwater and tailings fluids. If the total acidity of groundwater and contaminated sediments, as well as the neutralization capacity of the aquifer matrix are known, the distance that the acid plume will migrate can be estimated based on mass balance.

The acidity of tailings fluids and contaminated groundwater comes from the sulfuric acid in the processing of the ore and from hydrolyzable cations such as Al^{3+}, Fe^{3+}, and Mn^{2+} in the solution. Titratable acidity is often determined by the Solotto method (Solotto *et al.*, 1966). This procedure titrates an acid solution with NaOH to a pH of 7.3 and the total acidity is reported as $mg\,L^{-1}$ or $mol\,L^{-1}$ of $CaCO_3$ equivalent. However, laboratory titration results are unreliable because of the sluggishness of hydrolysis and precipitation reactions of Fe^{3+} (Hem, 1985). The end-point of pH 7.3 does not have a theoretical basis. Another common method is the "hot H_2O_2" method, which overcomes the kinetic problems (American Public Health Association, 1999).

The titration end-point of 8.3 does have a theoretical basis for calculating carbonate acidity, as discussed above and in Chapter 3. However, it becomes irrelevant in acid mine drainage waters. For the objective of projecting pH plume migration, it is both theoretically sound and practical to define an *operational acidity* as the amount of $CaCO_3$ needed to neutralize the water to a pH defined by saturation with calcite, at which point the fluid can no longer dissolve calcite. Geochemical modeling programs can be used to simulate this base-titration. The modeling results can be compared with laboratory titration. How good the modeling results are depends on our knowledge about the minerals allowed to precipitate at the initial conditions.

In this study, the geochemical modeling code PHREEQC, which has reaction path modeling capability, was used to calculate the moles of calcite needed to neutralize 1.0 liter of tailings pore fluid of TS-3 and contaminated groundwater of MW-86. The keyword REACTION was used to titrate calcite into the solution (Table 8.2). The mineral acidity of Al^{3+} and Fe^{3+} was taken into account by allowing amorphous $Fe(OH)_3$ and gibbsite to precipitate. Additionally, gypsum was also allowed to precipitate. The dissolution of the calcite produces CO_2, and it is assumed that CO_2 will escape when its fugacity is greater than 1 bar in this shallow unconfined aquifer. Therefore, we use the fixed fugacity reaction path model discussed in §2.3.2 for the simulation. All five phases have an input of 0.0 moles in the initial system so that they are not allowed to dissolve into the system.

Modeling Results

The results are shown in Figures 8.3 and 8.4. It will take dissolution of 0.125 and 0.04 moles of calcite to neutralize 1.0 liter of TS-3 and MW-86 water to a pH of 6.08 and 6.39, respectively, at which the solutions are saturated with calcite. At these points, no more calcite is dissolved into the solutions – titration of additional calcite into the system merely increases the mass of solid calcite. The equilibrium conditions concerning other constituents did not change from these points on. The amounts of precipitated gypsum do not increase.

```
TITLE Titrate TS-3 with calcite
SOLUTION 1    TS-3
        pH        3.8
        unit      mg/l
        temp      16.0
        Na        389.     charge
        K         60.
        Ca        310.
        Mg       1000.
        Cl        550.
        C(4)       5.
        S(6)   16500.
        Fe(3)  1950.
        Al     1020.
        Mn(2)    66.3
        Si       40.5
EQUILIBRIUM_PHASES 1
        Gibbsite        0.0   0.0
        Fe(OH)3(a)      0.0   0.0
        calcite         0.0   0.0
        gypsum          0.0   0.0
        CO2(g)
REACTION 1 Add calcite to the acid groundwater
        Calcite       1.0
        0.25    moles in 20 steps
SELECTED_OUTPUT
        -file    ts3.pun
        -a       H+
        -si calcite
        -equilibrium_phases  gypsum Fe(OH)3(a) gibbsite
END
```

Table 8.2. Titration of sample TS-3 with calcite. PHREEQC input file.

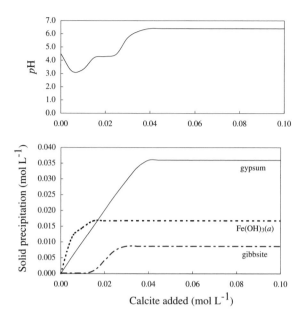

Figure 8.3. Results of titration of calcite into MW-86 water.

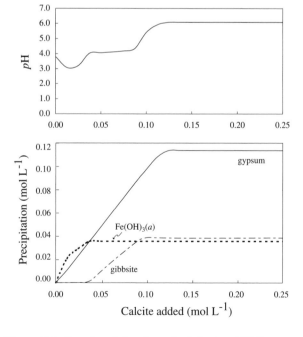

Figure 8.4. Results of titration of calcite into TS-3 water.

Note in Figures 8.3 and 8.4 that the solution pH first dropped after reacting with calcite because the initial solutions were supersaturated with respect to amorphous $Fe(OH)_3$. In the subsequent titration runs, we allow supersaturated $Fe(OH)_3(a)$ to precipitate, causing a drop in pH through the net reaction,

$$CaCO_3 + 1.5Fe(OH)^{2+} + SO_4^{2-} + 5H_2O = H_2CO_3^0 + 1.5Fe(OH)_3(a) + CaSO_4 \cdot 2H_2O + H^+$$

These modeling results can be compared with hand calculations. Drever (1988) showed calculations of mineral acidity in waters with high Al concentrations. Hedin *et al.* (1993) used the formula

$$\text{acidity} = 50 \left[\frac{2Fe^{2+}}{56} + \frac{3Fe^{3+}}{56} + \frac{3Al}{27} + \frac{2Mn^{2+}}{55} + 1000 \left(10^{-p\text{H}} \right) \right]$$

to calculate the acidity of acid mine drainages. The unit from this formula is $mg\,L^{-1}$ in equivalent $CaCO_3$. Using this formula, we calculated an acidity of 0.110 and 0.038 $mol\,L^{-1}$ $CaCO_3$ for TS-3 and MW-86, respectively.

Note that the formula above uses the "free" ions in calculations, ignoring the effects of speciation, and hence would overestimate the acidity. For example, our speciation modeling shows that the complex $Fe_3(OH)_4^{5+}$ is dominant. Precipitation of 1 mole of $Fe(OH)_3(a)$, therefore, only generates 1.6 moles of H^+, compared with 3 moles for Fe^{3+}.

Calculated acidity values are slightly smaller than our modeling results, however. This is because the formula ignores the acidity of carbonic acid. In our modeling set-up, the acidity from carbonic acid is not only from the initial solution but also from the calcite dissolved during the titration. When compared with a site that has numerous acidity measurements, Hedin *et al.* (1993) found that calculated acidity values are significantly smaller than those from the "hot H_2O_2" method in high acidity water.

8.4 pH Buffering

As we discussed in §6.2, the samples from the monitoring wells at the Bear Creek Uranium site are observed to fall into four fairly well defined pH zones (Figure 6.2). Compositional zones such as this are often produced by mineral buffering of the solutions. This process can be illustrated by using reaction path modeling.

We imagine a rock consisting of quartz, calcite ($CaCO_3$), goethite (or any form of ferric hydroxide, $Fe(OH)_3$), and kaolinite ($Al_2Si_2O_5(OH)_4$), to which we add a solution of sulfuric acid. The conceptual experiment is illustrated in Figure 8.5, and the results are shown in Figure 8.6. The PHREEQC file that produces these data is shown in Table 8.3.

As acid is added to the mineral assemblage, the pH stays relatively constant until calcite disappears. When calcite is gone, the pH drops to a new lower level, controlled by kaolinite. As the strength of the acid increases, eventually kaolinite also disappears, leaving only goethite, and the pH drops again to a level controlled by goethite. The dissolution of calcite raises the calcium content of the solution, causing gypsum to precipitate.

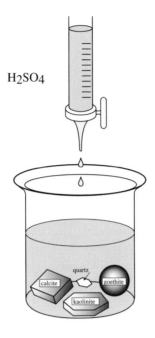

Figure 8.5. A conceptual experiment: A sulfuric acid solution is added to a rock consisting of goethite, kaolinite, calcite and quartz.

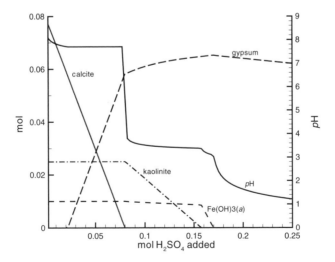

Figure 8.6. Quantities of minerals and acid during the titration of calcite, kaolinite, ferric hydroxide, and quartz with sulfuric acid.

```
TITLE Buffering example using titration
SOLUTION 1
        temp            25
        units           mol/kgw
        ph              7           charge
        Na              .1
        Cl              .1
EQUILIBRIUM_PHASES 1
        Quartz          0.6
        Kaolinite       0.0         .025
        Fe(OH)3(a)      0.0         .01
        Calcite         0.0         .08
        CO2(g)          -3.5
        O2(g)           -0.7
        Gypsum          0.0         0.0
INCREMENTAL_REACTIONS   true
REACTION 1
        H2SO4
        0.25 moles in 100 steps
SELECTED_OUTPUT
        -file           buf_tit
        -reset          false
        -step           true
        -ph             true
        -si             Fe(OH)3(a) Kaolinite Calcite Gypsum
        -equilibrium_phases Fe(OH)3(a) Kaolinite Calcite Gypsum
END
```

Table 8.3. PHREEQC input file for the buffering example.

Quartz is present throughout, and is relatively unaffected. In the presence of quartz, kaolinite is the expected aluminosilicate phase. However, as mentioned above, gibbsite ($Al(OH)_3$) or a gibbsite-like phase (such as one of its polymorphs, or an amorphous phase) is often found apparently to be controlling the dissolved Al concentration. Very similar buffering stages would be observed in this case.

The mineral proportions and acid volumes in Figure 8.6 are of no particular importance; they have been adjusted to illustrate the successive buffering that can result when one mineral after another is dissolved out of the rock.

8.5 Deep Well Injection of Hazardous Wastes

8.5.1 Background

The practice of employing deep injection wells to dispose of liquid waste started as early as the 1930s in the oil and gas industry. The fact that oil and gas have been trapped in deep geological reservoirs for millions of years prompted the idea that waste products might also be safely contained in these reservoirs. The practice of waste injection increased in the 1970s and 1980s, and has expanded from injecting oil-field brines to industrial wastes and treated sewage. In 1983, 11.5 billion gallons of hazardous waste were injected (US EPA, 1985). Most waste is aqueous effluent from chemical manufacturing. Waste streams can be very acidic (pH 0.03) or alkaline (pH 13.8), and contain organic compounds and heavy metals. Some are hazardous. The injection wells have an average depth of 4,000 feet (1219 m). About 62% of injection-well reservoirs are composed of sandstone, 34% of carbonates, and the remaining 4% of evaporites and shale (Warner and Lehr, 1977).

The US EPA is charged under the Safe Drinking Water Act of 1974 to promulgate regulations that protect current and potential underground sources of drinking water (USDW). USDW are defined as aquifers or portions thereof that supply any public water systems or contain water with <10 000 mg L^{-1} total dissolved solids. EPA's Underground Injection Control (UIC) program was established under this statute to ensure protection. The regulations establish five classes of wells.

Class I Injection of municipal or industrial waste below the deepest USDW.

Class II Injection related to oil and gas production.

Class II Injection for mineral recovery.

Class IV Injection of hazardous or radioactive waste into or above USDW.

Class V All other wells used for injection of fluids.

The enactment of the Resource Conservation and Recovery Act (RCRA) in 1984 resulted in EPA amendments to regulations in 1988, which prohibit underground injection of hazardous wastes, unless such injection is granted with an exemption from EPA (40 CFR 124, 144, 146, and 148). An exemption can only be granted by going through a process commonly known as "no-migration" demonstration or petition. The owners and operators must submit substantial information and demonstrate that migration outside of the injection zone will not occur until 10 000 years have lapsed. These demonstrations can be either based on "hydrological containment", which argues that waste would not reach the point of discharge within 10 000 years, or based on the chemical fate of the wastes, which argues that reactions in the formation have transformed the wastes so that they are non-hazardous.

The injection of waste streams that contrast greatly in chemistry from the host formation can cause severe geochemical reactions. These reactions have many important implications.

1. Geochemical reactions can change the hydraulic properties of the aquifer (formation damage) and therefore result in fluid migration to overlying USDW.

2. Precipitation, ion-exchange, and adsorption reactions can immobilize the waste constituents; and aqueous complexation, chelation, and colloid formation can mobilize others.

3. Geochemical reactions may transform the hazardous wastes into non-hazardous constituents or increase their toxicity by hydrolysis, neutralization, thermal and bio-degradation, oxidation–reduction, and other reactions.

4. Reactions cause a variety of operational problems (see the Box above).

However, only a few geochemical modeling studies have been used to assess these geochemical consequences (Coolby, 1971; Drez, 1988; Gardiner and Myers, 1992; Gunter *et al.*, 2000; Kharaka *et al.*, 1999). Smith (1996) reported that there have been only a few geochemical-fate petitions submitted to US EPA for review, and only one was accepted. Apps (1992) summarized the state-of-the-art of geochemical models and their suitability for applications to "no-migration" petitions based on chemical fate. Applications of geochemical modeling to waste-formation water compatibility, diagnosing well plugging and formation damage, and designing well-treatment methods are also not common. Although a number of difficulties in applying geochemical modeling to deep well injection have been identified, the lack of use of geochemical models has caused serious problems. Two well documented cases show that operators have injected waste streams that were already supersaturated with calcite and the precipitation of calcite caused well-plugging problems (see Bethke, 1996, and the case study below). These problems can be avoided if geochemical modeling is used as a screening tool. On the other hand, most applications have involved simple mixing calculations. More complex reaction path calculations such as flow-through and bio-reactor type reaction path calculations will help answer the questions of porosity changes over a spatial range, etc.

8.5.2   A Case Study

Zhu (unpublished data) used mixing calculations to evaluate the possible causes of well plugging at a deep-injection well facility operated by the Coastal Chemical Corporation near Cheyenne, Wyoming. The 6 000 ft deep wells are completed in the Hygiene

---

Geochemical reactions can result in many changes upon injection:

Formation damage.   Precipitation and dissolution can cause changes of permeability and caving. Such effects are no doubt in operation in the instances of injecting hydrochloric and sulfuric acids into carbonate aquifers or sandstone cemented by calcite.

Damage to the confining layer formation or sealed faults.   Dissolution reactions (e.g., dissolution of gypsum) can lead to the deterioration of confining formations. Surface adsorption of aluminum and ferric iron onto quartz can reduce the surface energy and cohesion, and breaking strength of the formation. These effects can result in the escape of fluids from injection zone to USDW.

Well blowout.   Injection of acid into carbonate formations will cause the dissolution of carbonate minerals and produce CO_2. When the pore pressure is high enough, fluids and wastes can be forced up the injection well and cause well blowout.

Well-casing failure.   Corrosive wastes can cause the failure of casing and packing and inadvertently inject wastes into other formations.

Well plugging.   Reduction of the permeability of the formation or well screens can be caused by precipitation, clay swelling as a result of dilution, or mobilization of fine particles.

Sandstone, which is composed of quartz, K-feldspar, clay, iron oxide, and calcite cement. Alkaline wastes from fertilizer manufacturing processes are injected into the formation. In 1996, the injectivity of the wells had decreased from 140 to 40 gallon per minute (gpm). Well plugging because of precipitation was suspected to be the cause. Well-treatment plans are subject to approval from the US EPA and state regulatory agencies. Geochemical modeling was called upon to evaluate the chemical compatibility of injectate with formation water and the causes for reduced permeability.

The Formation Water

There were no available complete chemical analyses of the formation water. Therefore, the chemical composition was inferred from two pieces of information: the water quality analysis dated 1969 and the mineralogy of the Hygiene Sandstone. The partial chemical analysis was conducted prior to injection and included pH, alkalinity, and chloride concentration (Table 8.4). In the model, the analytical alkalinity was as-

---

Apps (1992) pointed out serious deficiencies of geochemical models in assessing the effects of injecting hazardous wastes:

- The lack of thermodynamic data for aqueous organic species, organic-metal complexes, solid solutions, and surface adsorption; and kinetic data for heterogeneous reactions.

- The lack of the ability to model activity coefficients of concentrated mixed electrolytes.

- The lack of computer codes that include both transport and chemical reactions (coupled models).

- The lack of methods to evaluate uncertainties in modeling results.

However, should we just ignore geochemical reactions and not conduct any modeling altogether because we cannot model them accurately? Unfortunately, this appears to be the prevailing practice.

---

sumed to be HCO_3^-, and the solution was forced to be at equilibrium with calcite (so constraining the concentration of Ca^{2+}). The model also assumed equilibrium with K-feldspar, quartz, kaolinite, illite, and goethite. As a result, the concentrations of K^+, SiO_2, Al^{3+}, Mg^{2+}, and Fe^{3+} were calculated from their equilibria with the respective minerals.

The Injectate

An analysis of the waste pond water was used for calculating speciation and solubility of the injectate (Table 8.4). The modeling results show supersaturation of calcite and dolomite (and other carbonates), as well as iron oxides, quartz, and a number of unlikely silicates.

The results also show redox disequilibrium in the waste pond, which is not unusual. The analytical data show the co-existence of organic carbon, carbonate, nitrate, and ammonia in the pond. This has not been modeled in our example, because we did not feel it was an important part of the problem.[4]

Mixing Calculations

Chemical changes upon the mixing of injectate and formation water were simulated using PHREEQC (Parkhurst, 1995). PHREEQC calculates the changes of pH, temperature, speciation, and mineral precipitation and dissolution as a result of the mixing. In

---

[4]Note that to model redox disequilibrium in PHREEQC, a new species must be defined for each disequilibrium redox species (Parkhurst and Appelo, 1999, Example 15). Program REACT (The Geochemist's Workbench™) has a simpler *decouple* command.

```
DATABASE llnl.dat
TITLE Mixing injection fluid and formation water.

SOLUTION 1 Formation water.
        units           ppm
        pH              8.8
        temp            54.4
        Cl              129.
        Na              100.        charge
        Alkalinity      980.        as          HCO3
        N(5)            12.         as          NO3
EQUILIBRIUM_PHASES
        Calcite         0.0
        Quartz          0.0
        K-Feldspar      0.0
        Kaolinite       0.0
        Illite          0.0
        Goethite        0.0
SAVE SOLUTION 1
END
SOLUTION 2 Injection fluid.
        units           ppm
        pH              9.30
        temp            25.0
        Ca              55.
        Mg              18.
        Na              460.        charge
        K               12.
        Si              53
        N(-3)           2900.
        N(5)            1600.
        N(3)            14.
        Cl              320.
        F               1.6
        P               1.9
        Alkalinity      5800.
        C(-4)           460.
        S(6)            420.
        Ba              0.14
        B               6.2
        Fe(3)           0.14
        Sr              0.4
SAVE SOLUTION 2
END

EQUILIBRIUM_PHASES 1
        Calcite         0.2         0.0
        Quartz          0.0         0.0
        Goethite        0.0         0.0
MIX 2
        1           0.95
        2           0.05
SAVE SOLUTION 3
SAVE EQUILIBRIUM_PHASES 3
END
```

Table 8.4. PHREEQC input file for mixing fluids in a fertilizer problem. (Continued on next page.)

```
USE EQUILIBRIUM_PHASES 3
MIX 3
        3        .95
        2        .05
SAVE SOLUTION 4
SAVE EQUILIBRIUM_PHASES 4
END
USE EQUILIBRIUM_PHASES 4
MIX 4
        4        .95
        2        .05
SAVE SOLUTION 5
SAVE EQUILIBRIUM_PHASES 5
END

===================== MIX 5 and MIX 6 steps omitted here ===========

USE EQUILIBRIUM_PHASES 7
MIX 7
        7        .95
        2        .05
END
```

Table 8.4. (Continued from previous page.) PHREEQC input file for mixing fluids in a fertilizer problem.

the model, we specified that calcite, quartz, and goethite can be precipitated from the mixture when they are supersaturated at SI > 0.2, 0.0, and 0.0 for calcite, goethite, and quartz, respectively. The mixing calculations were carried out using fractions of 0.95 and 0.05 for the formation water and injectate, respectively. The mixed solution is saved and mixed with another 5% injectate sequentially. After six of these mixing operations (Mix No. 7) the solution has about 74% formation fluid and 26% injectate.

The results show that calcite and goethite are precipitated from the solution (Figure 8.7). Calcite has the largest amount of precipitation, $0.41 \, \mathrm{mmol \, kg^{-1}}$ upon mixing of 5% injectate six times with the formation water. The amounts of precipitation increase linearly at increasing injectate fractions. This relationship may indicate that calcite precipitation does not necessarily result from the mixing of the two waters with different chemistry, but from the fact that we allow calcite to precipitate from the injectate supersaturated with calcite (SI = 2.19). Calcite has a retrograde solubility, and an increase in temperature will increase the potential for precipitation.

The geochemical modeling results suggest that calcite precipitation is the major cause of the reduced permeability. For temporary relief, precipitated calcite can be dissolved with a dilute hydrochloric acid (HCl). Past application of HCl typically improved the injectivity within 48 hours. However, regular use of the acid would probably result in significant formation damage because of the subsequent mobilization of carbonate from the well bore into the formation. Also, regular acidification would be likely to damage the perforated casing.

A long term solution to the problem would need to deal with the supersaturation of calcite in the waste stream. If pond water is not pre-treated, calcite will continuously

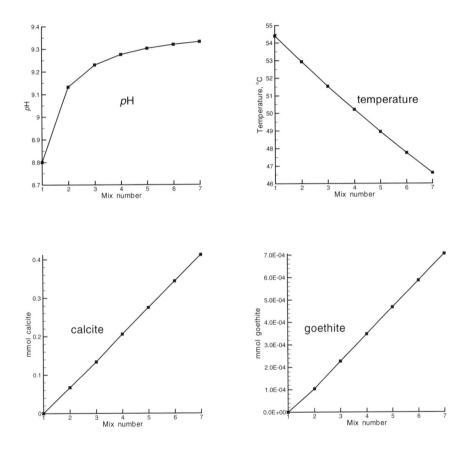

Figure 8.7. Results of mixing calculation for deep well injection. Mineral concentrations are given in millimoles per liter.

precipitate at areas close to the well bore upon injection. Pre-treatment of pond water before injection may provide a solution to the problem.

Note that, in this study, only the mixing of formation water and waste stream was simulated. The formation matrix (minerals in the sandstone) was not part of the simulated groundwater chemical system. Different kinds of models need to be constructed if the solid matrix is included as part of the system (see §8.1).

Comments

This study demonstrates the potential use of geochemical modeling in evaluating operational problems and waste-formation water compatibility associated with the injection of hazardous wastes into deep wells. Rather, it demonstrates that the failure of conducting the simplest geochemical modeling can result in problems that can otherwise be avoided. Bethke (1996, p. 309) gives another example of geochemical modeling application to deep well injection in Illinois, and vividly demonstrates a design failure because the design did not consider geochemical compatibility of the formation and wastes. Modeling of the geochemical reactions involving wastes of complex chemical compositions and deep saline aquifers is difficult because of the many geochemical processes involved, the lack of thermodynamic data, and deficiencies of modeling codes (Apps, 1992; Boulding, 1990). However, even given the possible errors and simplifications of geochemical modeling, some obvious and costly mistakes can be avoided.

8.6   Pit Lake Chemistry

Davis and Ashenberg (1989) used PHREEQE (a forerunner to PHREEQC) to perform titration and mixing calculations to evaluate the option of neutralizing Berkeley Pit water with alkaline tailings fluids. The Berkeley Pit Superfund site is located in the Butte Mining District of Montana (Figure 8.8). Flooding of the pit, starting in 1983, with acidic water of high concentrations of Fe, SO_4^{2-}, Al, As, Cu, Cd, Cu, K, Mg, Mn, Na, and Zn caused the water level to rise rapidly at 22 m per year. If no remediation actions are implemented, the pit lake water will spill over in the year 2009, and, before that time, there is potential for encroachment of contaminated water into the alluvial aquifer and the transport of metals into Silver Bow Creek. Therefore, the Berkeley Pit has been under intensive study. In addition to the local environmental concerns, the Berkeley Pit also serves as a good study ground for pit lake chemistry in general, because most pit lakes from stripping mining operations have yet to be filled.

The deepest part of the pit lake is 542 m, and the lake extends laterally 1.8 km by 1.4 km (Figure 8.8). Flooding started one year following the cessation of de-watering pumpage in the adjacent Kelley Shaft in April 1982. Inflow to the pit includes surface water runoff and groundwater inflow due to seepage from the Yankee Doodle tailings pond, the leachpad, and the West Camp Mines. Davis and Ashenberg (1989) first studied the chemical profile of the water body in the Berkeley Pit. They found that the chemistry at the top of the water body is different from that in the deep part of the lake, and they attributed the differences to mixing with surface runoff and seasonal chemical reactions. By analyzing the chemical compositions of the particulate on the filters and calculation

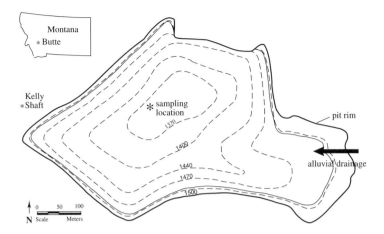

Figure 8.8. Location and depth contours of the Berkeley Pit Lake.

Condition	Case	Minerals allowed to precipitate and dissolve
No CO_2 exchange with atmosphere	1	none
	2	gypsum, $Fe(OH)_3(a)$
Infinite CO_2 reservoir with P_{CO_2} of $10^{-3.5}$	3	gypsum, $Fe(OH)_3(a)$
	4	gypsum, $Fe(OH)_3(a)$, jurbanite
	5	gypsum, $Fe(OH)_3(a)$, jurbanite, jarosite

Table 8.5. Five modeling scenarios simulated by Davis and Ashenberg (1989).

of the SI for water samples, they concluded that gypsum, ferric hydroxide, jarosite, and jurbanite are the solids controlling metal solubilities.

Davis and Ashenberg (1989) used titration and mixing calculations to evaluate the remedial action of using alkaline tailings fluids from the mineral processing operations. They performed the simulation by titration of the tailings fluid into the pit water from the oxidized zone in the top few meters of the lake. Five modeling scenarios were simulated. The first two cases simulated reactions in a system closed to the atmosphere, to mimic winter conditions when the lake is covered with ice. Cases 3 to 5 simulated the reactions in a system open to the atmosphere with a P_{CO_2} of $10^{-3.5}$. (This is the reaction path configuration with fixed fugacity; see §2.3.2 and also Stumm and Morgan (1996), pp. 157–160 for reference.) Minerals were introduced into the modeling scenarios sequentially to see the effects of precipitation on the neutralization of pit lake water. Table 8.5 shows the conditions of the five cases.

Note that reactions that involve the precipitation and dissolution of the minerals, with the exception of gypsum, in Davis and Ashenberg's (1989) simulations consume or produce protons. Therefore, they more or less buffer the pH changes upon the mixing of the pH 11.3 tailings fluid with the pH 2.7 pit lake water. It should also be noted that equilibrium with atmospheric CO_2 also constitutes a buffer reaction.

The results of the simulation are presented in Figure 8.9, and the following can be

Solid or gas phases	Reaction	$\log K^a$
$Fe(OH)_3(a)$	$Fe^{3+} + 3\,H_2O = Fe(OH)_3 + 3\,H^+$	-15.7
Jurbanite	$Al^{3+} + SO_4^{2-} + H_2O = AlOHSO_4 + H^+$	-3.23
Gypsum	$Ca^{2+} + SO_4^{2-} + 2\,H_2O = CaSO_4 \cdot 2H_2O$	-4.85

^{a}Data from PHREEQE, with the exception of $Fe(OH)_3(a)$ and $CO_2(g)$.

Table 8.6. Chemical reactions used by Davis and Ashenberg (1989).

inferred from Davis and Ashenberg's (1989) graph.

Case 1   Rapid increase of pH occurs with increasing amounts of alkaline tailings fluids mixed with acidic pit water. This is an unrealistic scenario because Fe and Al in the acidic pit lake water will precipitate and buffer pH in one solid form or another.

Case 2   This case should be compared with Case 1 to examine the effects of solid precipitation reaction to buffer the pH of mixed fluids. Slow increase of pH from 2.7 to 3.5 is caused by buffering of $Fe(OH)_3$ (see Table 8.6). Rapid increase of pH from 3.5 to 5.5 is due to lack of buffering reactions. It should be noted that if the authors had allowed $Al(OH)_3$ or kaolinite to precipitate in their simulations, the pH of the mixtures would be buffered at about 4.5 until most Al is precipitated from the solutions (see Section 6.2). The authors measured Al concentrations of about 200 $mg\,L^{-1}$ in the pit lake.

Case 3   This case should be compared with Case 2 to see the effect of buffering with an external reservoir of CO_2 gas. The CO_2 gas dissolution should have a buffering effect on the pH in the range of about pH > 6.3. The pH stabilizes at 8.0.

Case 4   This case should be compared with Case 3 to see the effect of jurbanite precipitation. pH stablizes at 7.0.

Case 5   Compare with Case 4 to see the effect of jarosite precipitation. pH stabilizes at 5.6.

   According to Davis and Ashenberg (1989), the results of the simulations provide some useful information for the evaluation of the feasibility of using alkaline tailings fluid to neutralize Berkeley Pit water as a remedial action. The authors calculated that it needs three volumes of alkaline fluids to neutralize the pit lake water to pH of 5.0 with the Case 3 scenario. That would result in a water level 20 m above the contact between the alluvial aquifer and the subjacent bedrock, assuming all inflows into the pit except alluvial and bedrock groundwater contributions can be controlled. The authors concluded that this remedial alternative is a "tractable" solution to the acid mine drainage problems at the Berkeley Pit.

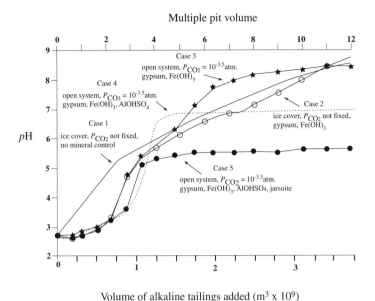

Figure 8.9. pH of the Berkeley Pit lake as a result of the mixing with an alkaline tailings fluid.

8.7 Artificial Recharge

Artificial recharge into underground formations has been applied to replenish depleted supplies, prevent or retard salt water intrusion, or store water underground (US EPA, 1999). Recharge methods range from surface infiltration to injection wells. Water used for artificial recharge includes surface water, stored flood water, treated municipal waste water, and so on. The recharged water may not be geochemically compatible with groundwater in the formation, and geochemical reactions may adversely affect the operations or the water quality.

As a result of these concerns, geochemical modeling was performed to evaluate the consequences of artificial recharge into deep aquifers in the Denver basin (Ring *et al.*, 1986), the Albuquerque basin and the El Paso basin (Whitworth, 1995), the Las Vegas basin (Brothers and Katzer, 1987), and the dunes of the Amsterdam water supply (van Breukelen *et al.*, 1998). Modeling was either used as a screening tool (Whitworth, 1995) or as a tool to aid interpretation of a demonstration experiment (Brothers and Katzer, 1987). Geochemical evaluation of the feasibility of the projects include

1. whether the mixing of the recharge and groundwater will cause precipitation and, therefore, clog the wells and reduce the operational life of the project; and

2. whether the artificial recharge will have an adverse impact on the groundwater quality.

Despite the obvious benefits, not many geochemical modeling applications to artificial recharge problems are in the open literature. We have found no suitable example here.

8.8   Applications to Natural Background Studies

Natural background or baseline concentrations of metals in surface and groundwater refer to those concentrations prior to disturbance by human activities. In the mineralized areas, the occurrences of low pH and high metal concentrations in groundwater and surface water may result from natural weathering and leaching of exposed ore bodies and wall rocks. In such cases, it may be technically or economically impossible to meet remediation standards that are below natural background concentrations (e.g., drinking water standards – MCLs; or Aquatic Water Quality Criteria –AWQC). Therefore, alternative remediation targets should be used, which should reflect natural background concentrations. On the other hand, in many areas, mining and mineral processing practices or other forms of land use have significantly accelerated the leaching and weathering processes. Nordstrom *et al.* (1996) contend that at Iron Mountain, California, mining activities have accelerated the acid generation rate by 1000 times the natural weathering processes. In general, natural background should not be used as an excuse for escaping clean-up responsibilities.

Often, environmental concerns and clean-up activities began after centuries of mining activities and land uses. It is quite difficult, therefore, to establish the natural background before human interference. Several methods are useful for establishing background concentrations, such as examination of historical records, comparison between mined and un-mined mineralized areas (natural analogs), conduction of baseline geochemistry studies prior to mining operations, and leaching experiments (Runnells *et al.*, 1992). Frequently, these approaches are not applicable to some sites, and geochemical modeling can be helpful in establishing the range of metal concentrations.

Runnells *et al.* (1992) used PHREEQE to simulate the possible ranges of metal concentrations in surface and groundwater at the Tri-state Zn–Pb mining district, Kansas. The ore deposits at the Tri-state district mainly consist of Zn, Pb and Fe sulfides hosted in carbonate rocks. Runnells *et al.* (1992) started by choosing three mineral assemblages that they believe are representative of highly oxidized, moderately oxidized, and slightly oxidized ores. They also selected Eh and P_{CO_2} conditions that they believe represent surface water or shallow groundwater and deep groundwater. Then, they let PHREEQE calculate the equilibrium concentrations of metals in pure water and upstream uncontaminated water in the area. The resultant metal concentrations are controlled by the solubility of the mineral phases they selected and under the chosen P_{CO_2} and Eh conditions. They then compared the simulated metal concentrations with those of the contaminated section of western Short Creek. They concluded that natural weathering and leaching are capable of producing the observed high metal concentrations.

The approach to calculating ionic compositions of the aqueous solution from solubility of the minerals at equilibrium under certain T, P, pH, Eh, and gas fugacity conditions is a well established method in geochemistry (e.g., Anderson and Crerar, 1993; Zhu *et al.*, 1994). Runnells *et al.* (1992) attempted to apply it to natural background studies. Their models are nevertheless simplistic, as the authors admitted in the paper. The choices of the mineral phases are critical because the metal concentrations are calculated from their solubility. The calculations of aqueous fluid compositions in petrological and ore deposit studies rely on careful characterization of mineralogical identifications and petrographic examination. There is no reason that such studies should be absent when this modeling approach is applied to environmental studies. Furthermore, under surficial conditions, metals with low concentrations are often not controlled by pure solid phases in which they are major components, but by co-precipitation, ion-exchange, and surface adsorption. Hence, the natural background concentrations may be elevated by natural weathering and leaching, but may also be lowered by the natural attenuation processes.

Recognizing the limitations of this modeling approach, the US EPA responded to ASARCO's modeled background concentrations for the California Gulch site in Colorado, stating "The metal values predicted by ASARCO could at best only be used to indicate upper limits for a few localized pockets of slow moving ground water" (Record of Decision, 1988, COD980717938). However, more complex geochemical models may provide more realistic estimates of natural background geochemistry. Nordstrom *et al.* (1996) have pointed out that geochemical modeling of natural background may expand the scope and include inverse mass balance models, reaction path models, and models with kinetics. We would add models with surface adsorption, ion-exchange, and reactive transport to the list of potential applications.

9

Inverse Mass Balance Modeling

9.1   Introduction

Garrels and Mackenzie (1967) introduced inverse mass balance modeling into geochemistry. They showed that if the chemistry of the start and end solutions are known, possible mass transfer reactions that had produced the compositional differences and the extent to which these reactions had taken place could be deduced from the mass balance principle.

Plummer and co-workers (Plummer, 1985; Plummer *et al.*, 1983, 1990, 1991, 1994; Wigley *et al.*, 1978) used this concept to model groundwater aquifers, and greatly expanded and formalized this approach. Their main interest was to deduce the mass transfer reactions taking place between two observation points along a flow path, which may have been responsible for the chemical and isotopic evolution of the groundwater. Mass transfer here refers to simple mass transfer between two or more phases, such as dissolution and precipitation of minerals (e.g., Nordstrom and Munoz, 1985). The development of the program NETPATH by Plummer and co-workers (Plummer *et al.*, 1991, 1994) has greatly facilitated the use of this modeling approach. Recent development of a new version of PHREEQC by David Parkhurst (Parkhurst, 1995, 1997) incorporates uncertainty analysis and a more complete set of mass balance constraints, reaching a new level of model sophistication.

We omit the mathematical development here. Serious modelers should read Parkhurst (1997), Plummer *et al.* (1991, 1994), and Wigley *et al.* (1978). Readers are also encouraged to read very carefully the work by Plummer *et al.* (1990) on the Madison aquifer, Montana, which demonstrates splendidly the application of inverse mass balance modeling in a regional aquifer. The manuals for NETPATH and PHREEQC contain detailed tutorials for the programs with examples. The examples given here emphasize environmental applications in the regulatory environment.

9.2   Model Assumptions

The applicability of inverse mass balance modeling hinges on a number of assumptions, which are seldom examined in detail:

1. the two water analyses from the "initial" and "final" wells should represent packets of water that flow along the same path;

2. dispersion and diffusion do not significantly affect solution chemistry;

3. a chemical steady state prevailed during the time considered;

4. the mineral phases used in the calculation are or were present in the aquifer.

   The first requirement, that the final water must be evolved from the initial water, is obvious. The question is how to assure that this is the case. Often a geochemical modeling study is launched because the hydrogeologists in charge of the project have difficulty figuring out flow directions from hydraulic data alone. The ideal way, perhaps, is for the hydrogeologists and geochemists to work closely together and to use an iterative approach. If sufficient hydraulic data are available, particle tracking and mass transport simulations are good ways to deduce flow paths (e.g., Voss and Wood, 1994; Zhu *et al.*, 1998).

   Part of the compositional differences in the initial and final waters may result from hydrodynamic dispersion, a process that tends to spread the concentrations due to fluid mixing (Fetter, 1993; Freeze and Cherry, 1979). If not accounted for otherwise, inverse models implicitly attribute dispersion effects to mass transfer reactions (Plummer *et al.*, 1992; Zhu and Murphy, 2000). This assumption severely limits the applicability of inverse mass balance modeling. Fortunately, in large regional aquifers where lithology is relatively homogeneous and the time-frame of interest is long, hydrodynamic dispersion has negligible effects on concentration distributions (Johnson and De Paolo, 1996; Phillips *et al.,* 1989; Wigley *et al.*, 1978, appendix A; Zhu, 2000; Zhu *et al.*, 1998). Numerical transport simulation is a good tool to test this assumption. Zhu *et al.* (1998) used various values of dispersivity in their study of the Black Mesa basin, Arizona, and found that dispersion has negligible effects on ^{14}C transport for simulations of tens of thousands of years.

   Hydrodynamic dispersion may however be significant in small, local hydrogeological problems, such as a point source contamination (Plummer *et al.*, 1992). Another instance where diffusion may play an important role in water chemistry is the diffusion from permeable to less permeable parts of the aquifer, or matrix diffusion. This process appears to be important in fractured aquifers (Maloszewski and Zuber, 1991; Neretnieks, 1981), volcanic rock aquifers, aquifers adjacent to confining units (Sudicky and Frind, 1981), and sand layers inter-stratified with confining clay layers (Sanford, 1997). In systems in which a chemical steady state (see below) has not been reached, matrix diffusion effects may severely limit the applicability of inverse mass balance modeling to those systems.

   A chemical steady state is a sufficient, although not necessary, assumption because the underlying mathematical equations to be solved in inverse modeling are the integrated forms of mass conservation (Lichtner, 1996, p. 61). In other words, we have chemical analyses of the "initial" and "final" water, typically taken at approximately the same time, and now we want to figure out the mass transfer reactions between the two points from A (initial) to B (final). It takes x years for the packet of water to travel from A to B. It is necessary that the chemical composition at A at the time of analysis has not changed in x years and therefore represents the chemical composition of the packet of water from which B has evolved.

From this assumption, inverse mass balance modeling is more likely to be applicable to large regional aquifer systems, but less applicable to aquifers with point source contamination. For example, in the acid mine drainage impacted aquifer described in Chapter 6, the chemistry at all points is changing with time. Applicability of inverse mass balance modeling to this type of system depends on the spatial locations of the flow path and the time-frame of interest.

When the necessary condition that the chemistry at point A does not change with time is met, inverse mass balance models are still applicable when the chemistry at point B changes with time. An example is a laboratory column study, in which the chemistry of influent is maintained in the experiments while the effluent chemistry continues to change. In this case, we are assured that the effluent is chemically evolved from the influent. The variation of chemistry with time in the effluent does not violate the steady-state assumption. Another example is field injection of reactive tracers, during which the injectate chemistry is constant. Actually, laboratory titration experiments would also fit into this category because we know the initial solution chemistry from which the final solution evolves. Inverse mass balance modeling should find applications in these situations.

Therefore, for field applications, a chemical steady state is implicitly assumed in nearly all inverse mass balance modeling applications, but this has rarely been examined in detail. However, recent findings of large variations of groundwater flow patterns due to paleoclimate changes in the late Wisconsin and Holocene (Zhu *et al.*, 1998) raises doubts about this necessary assumption for inverse mass balance modeling in large regional aquifers.

Steady states for flow and for chemistry are different concepts and are often confused: steady-state flow means constant directions and magnitudes of velocities; steady-state chemistry means constant concentration distributions. Local equilibrium can possibly maintain a chemical steady state for reactive constituents in a transient flow regime (e.g., Ca^{2+} and HCO_3^- in a Karst aquifer). However, if kinetics plays a significant role, e.g., the dissolution rate of feldspars or oxidation of organic carbon, then the chemical state in a transient flow field depends on competing factors of kinetic rates and velocity changes. Thus, chemical steady state may be achieved for some constituents but not others.

Regarding the fourth assumption, solute-flux-derived mass transfer models are non-unique. Mathematically, this is an inversion problem. There are many different combinations among possible mineral and gas phases that can produce the same compositional differences in the initial and final water. Plummer *et al.* (1992) comment

> The validity of the mass-balance models depends significantly on the geo-chemical insight of the modeler in selecting appropriate phases in the model.

A number of considerations may help to eliminate unrealistic model solutions:

- Not all phases are present or were probably present in the aquifer. Model solutions that require phases unlikely to have ever been present in the aquifer are not realistic. Careful petrographic observation and mineralogy studies are the most direct evidence for mass transfer reactions, but are not commonly employed in modeling studies.

- Calculation of Saturation Indices from solution chemistry can also be a useful elimination tool (Plummer *et al.*, 1991). Models that precipitate minerals that are undersaturated and dissolve minerals that are supersaturated may not be realistic. The modelers must be cautious, however, in using this criterion. Inverse modeling calculates the "net" mass changes along a flow path, often a few kilometers apart. It does not consider the point to point mass transfer or equilibrium state along the flow path. A mineral phase may precipitate in one segment of the path, but dissolve in another.

- Chemical kinetic considerations are another criterion. For example, a model can be eliminated if it requires the precipitation of a mineral that is known not to precipitate readily in low temperature systems.

- Isotopes may help to eliminate unrealistic solutions by providing additional constraints (Plummer *et al.*, 1991, 1992, 1994). However, alternative interpretations of isotope data are often possible, and our lives do not necessarily become easier.

- Finally, Parkhurst (1997) found that inclusion of uncertainty analysis can reduce the number of phases needed to reproduce the net chemical changes and thereby reduce the number of models.

9.3 Groundwater Genesis, Black Mesa, Arizona

The Navajo sandstone in northeastern Arizona produces pristine groundwater and is the most productive unit of a regional aquifer, referred to as the N aquifer in the area (Cooley *et al.*, 1969). The N aquifer is the principal source for municipal, domestic, industrial, livestock, and agricultural water in this arid to semi-arid area (Figure 9.1). Withdrawal from the N aquifer has increased six-fold from 1967 to 1992 (Littin, 1992). Long term environmental sustainability is a grave concern in this area. Increased pumpage may reduce flow from sacred springs and degrade water quality by inducing leakage from an overlying saline aquifer. Numerical groundwater flow models have been developed as a water resource management tool (Brown and Eychaner, 1988; Eychaner, 1983; GeoTrans, Inc., 1987b). However, predictions from numerical models depend on the parameters used in the model. In particular, the recharge to the aquifer is a major component of the water budget, and predicted environmental impacts depend on the recharge values specified in the model. However, recharge in the arid and semi-arid areas is extremely difficult to estimate.

The main recharge area to the N aquifer is located in the northwestern area of the Mesa, near Shonto (Cooley *et al.*, 1969). Recharge in this area is mostly from snow

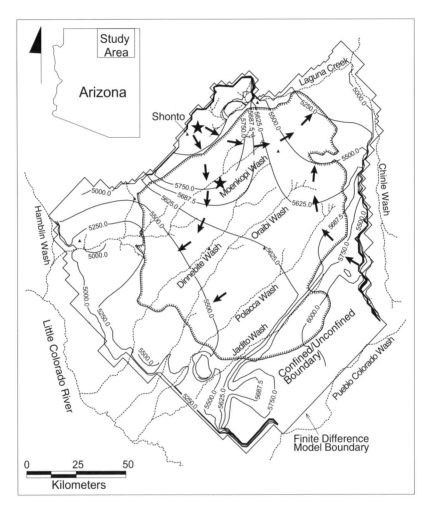

Figure 9.1. Simulated contours of present-day hydraulic heads (in feet) generated from the calibrated two-dimensional finite difference flow model developed by GeoTrans (1987b). Arrows indicate the directions of groundwater flow. Also shown are the finite difference model and the confined/unconfined aquifer boundaries. Star symbols indicate the two water samples used in inverse modeling.

Component	Na	Ca	Mg	K	C	SO$_4$	Cl	SiO$_2$
Initial concentration	0.151	0.848	0.109	0.0274	2.1004	0.0284	0.0762	0.0933
Snow	0.0261	0	0.0082	0.0153	0.0252	0.0167	0.0056	0
Minus snow	0.1249	0.848	0.1008	0.0121	2.0752	0.0117	0.0706	0.0933
Remove gypsum and halite	0.0543	0.8363	0.1008	0.0121	2.0752	0	0	0.0933
Change kaolinite back into plagioclase with all available SiO$_2$	0.00765	0.8077	0.1008	0.0121	2.0752	0	0	0
Change excess Na$^+$ to Ca–Na exchange and excess K$^+$ to Mg–K exchange	0	0.8115	0.1069	0	2.0752	0	0	0
Remove Ca^{2+} and Mg^{2+} back to calcite and dolomite	0	0	0	0	1.1568	0	0	0

Table 9.1. Inverse mass balance calculation. Concentrations are millimoles per liter.

melt in the winter and spring. It is necessary to understand the genesis of groundwater in the aquifer in order to make ^{14}C age corrections (see §9.5). Zhu (2000) performed inverse mass balance modeling from snow melt to wells in the Shonto area.

The inverse modeling calculations followed Garrels and Mackenzie (1967). Here, we show the example calculation for the well Navajo National Monument. Unlike Garrels and Mackenzie, who most likely "used back of the envelope" calculations, we used the Excel™ spreadsheet. Our calculation procedures, however, almost exactly followed Garrels and Mackenzie (1967).

The components Na, Ca, Mg, K, C, S, Cl, and Si were used to reconstruct the sources of groundwater constituents. The average chemical concentrations of eight snow samples analyzed by Dulaney (1989) were first subtracted from the groundwater concentrations to derive the materials from water–rock interactions (Table 9.1). Then, Cl and S were assigned to halite and gypsum, respectively (the row "remove gypsum and halite" in Table 9.1). Next, Na, Ca, C, and Si were subtracted in proportions required to turn all available silica back into a plagioclase through kaolinite precipitation. The plagioclase was assumed to have a composition of NaO$_{.62}$CaO$_{.38}$Al$_{1.38}$Si$_{2.62}$O$_8$. A total of 0.0752 moles of plagioclase was found to be needed through the reaction

$$0.0286\,\mathrm{Ca}^{2+} + 0.0467\,\mathrm{Na}^+ + 0.0519\,\mathrm{Al_2Si_2O_5(OH)_4} + 0.0933\,\mathrm{SiO_2} + 0.7058\,\mathrm{HCO_3^-}$$
$$= 0.0752\,\mathrm{Ca_{.38}Na_{.62}Al_{1.38}Si_{2.62}O_8} + 0.7058\,\mathrm{CO_2} + 0.4567\,\mathrm{H_2O} \quad (9.1)$$

This left "excess" Na and K, which were then attributed to Ca–Na, and Mg–K exchange. The former was found to be an important reaction in the N aquifer (Wickham, 1992); the latter reaction was arbitrarily assigned, but should be identical to Ca–K exchange or dissolution of sylvite in terms of calculating mineral carbonate. After these calculations, considerable amounts of Ca and Mg still remained, and were then assigned to calcite and dolomite. The carbon left unaccounted was then assumed to be from the dissolution of soil CO$_2$ gas. These calculations indicate that about 44 % carbon in the groundwater sample is from mineral sources and 56 % is from soil gas.

These spreadsheet calculations are very intuitive; they helped in the understanding of inverse mass balance modeling. We strongly recommend readers to practice using a spreadsheet at least once.

The same sort of calculations can now be carried out by using NETPATH, which is an interactive code. The power of a computer code is that one can repeat similar calculations and evaluate alternative mass transfer models with ease. For example, one can specify many phases, more phases than components, which may result in many different models.

For this NETPATH calculation, we specified the same eight components and the same nine phases in the input file as used in the spreadsheet calculation. NETPATH produced one model (Table 9.2). In the input file, we forced kaolinite to precipitate only and dolomite and calcite to dissolve only, and all three phases must be included in every model. If we do not include these constraints, more models will be produced. Inspection of model results shows that NETPATH produces results close to our spreadsheet calculations. However, the NETPATH model does not include Na–Ca exchange.

One can also use PHREEQC to perform inverse mass balance modeling for the same calculations. PHREEQC is capable of taking into account the analytical uncertainties. Here, we assume the analyses of snow and groundwater samples carry $\pm 6\%$ errors. The input file is listed in Table 9.3. The readers are referred to the PHREEQC manual for a detailed explanation of the input file (Parkhurst and Appelo, 1999).

Five different models are produced by PHREEQC, and they are compared with each other as well as with the spreadsheet calculation in Table 9.4. The major differences in the five computer calculated models are in ion-exchange reactions. In PHREEQC, Na, K, Mg, and Ca are allowed to exchange with each other, while in the spreadsheet calculation we only allowed Na–Ca and K–Mg exchanges. The PHREEQC models are slightly different from the spreadsheet calculation because of a 6% upward adjustment of the total carbon in PHREEQC.

We mentioned earlier that inverse models produce non-unique solutions, and in the absence of an excellent petrographic study, the choice of phases and hence the solutions appear to be infinite and arbitrary. We have to bear in mind that there are always inherent uncertainties in our models, and we must decide whether the results are useful despite the assumptions and uncertainties.

For example, in the preceding calculations, if we assume magnesite rather than dolomite is the carbonate phase containing Mg, we will have a different model that also satisfies the compositional difference in the snow and groundwater. If we choose K–Ca exchange rather than K–Mg exchange to account for the excess K, we would have another model. However, these choices give exactly the same results in terms of the proportions of mineral versus soil CO_2 derived DIC (dissolved inorganic carbon) in groundwater. One may strongly disagree with the choice of ion-exchange reactions. However, all the five models produced by PHREEQC (Table 9.4) give exactly the same results — 61% dissolved inorganic carbon is from soil CO_2.

Of course, Mg can also come from dissolution of biotite or amphiboles. Similarly, K can come from the dissolution of orthoclase, micas, and amphiboles. The models can become very complicated, and we never can be sure what exactly are the mass transfer reactions that produced the groundwater chemistry. It is an ambitious proposition that we can know the groundwater genesis. Rather, inverse mass balance modeling is a useful tool that can help us to understand it. Sometimes, the results can be used quantitatively, and are sufficient for the purpose of our studies.

```
Initial Well : snow
Final Well   : navajo nat monu

          Final      Initial
C        2.1004       .0252
S         .0284       .0061
CA        .8485       .0217
NA        .1510       .0126
SI        .0937       .0000
MG        .1090       .0029
CL        .0762       .0082
K         .0274       .0054


KAOLINITAL  2.0000SI  2.0000
DOLOMITECA  1.0000MG  1.0000C   2.0000RS  8.0000I1   .0000I2   .0000
CALCITE CA  1.0000C   1.0000RS  4.0000I1 -3.2000I2 71.0000
CO2 GAS C   1.0000RS  4.0000I1-19.9000I2100.0000
PLAGAN38CA   .3800NA   .6200AL  1.3800SI  2.6200
nacaexchCA  1.0000NA -2.0000
NaCl     NA  1.0000CL  1.0000
GYPSUM   CA  1.0000S   1.0000RS  6.0000I3 22.0000
mg/k excMG  1.0000K  -2.0000



           MODEL   1
  KAOLINIT  - F       -.10180
  DOLOMITE  + F        .11714
  CALCITE             .64421
  CO2 GAS            1.19666
  PLAGAN38  +         .11348
  NaCl                .06799
  GYPSUM              .02228
  mg/k exc           -.01100
```

Table 9.2. Output from NETPATH calculations for well Navajo National Monument water.

9.4   Acid Mine Drainage, Pinal Creek, Arizona

The Pinal Creek basin in central Arizona, about 100 km east of Phoenix, is one of the most polluted acid mine drainage sites in the USA (Figure 9.2). Mining activities over a century, mostly in copper from porphyry deposits, have resulted in a plume of acidic groundwater in the alluvium and basin fill aquifer. The plume is laden with heavy metals. Many potential contamination sources, including mine tailings, heap leach, surface impounds, and surface run-off and groundwater flow in the mineralized

```
TITLE SOLUTION 1 snow
    units   mmol/l
    Ca      0.0
    Mg      0.0082
    Na      0.0261
    K       0.0153
    Cl      0.0056
    S(6)        0.0167
    Si      0.0
    C       0.0252
SOLUTION 2 Navajo National Monument well
    units   mmol/l
    Ca      0.848
    Mg      0.109
    Na      0.151
    K       0.0274
    Cl      0.0762
    S(6)    0.0284
    Si      0.0933
    C       2.1004
INVERSE_MODELING
    -solutions 1 2
    -uncertainty 0.06
    -phases
        calcite     dis
        dolomite    dis
        gypsum      dis
        An38
        kaolinite
        CO2(g)
        halite
        NaX
        CaX2        force pre
        MgX2        force pre
        KX
    -range
PHASES An38
    Ca.38Na.62Al1.38Si2.62O8 + 5.52H+ + 2.48H2O
            = 0.62Na+ + 0.38Ca+2 + 1.38Al+3 + 2.62H4SiO4
        log_k 0.0  #No log_k, Inverse modeling only
END
```

Table 9.3. Input file for PHREEQC calculation for well Navajo National Monument water.

Models	Calcite	Dolomite	Gypsum	An$_{38}$	Kaolinite	CO$_2$	Halite	NaX	CaX2	MgX2	KX
Hand	0.7046	0.1069	0.0117	0.0752	-0.0519	1.1568	0.0706	0.0076	-0.0038	-0.0061	0.0121
1	0.6706	0.0943	0.0126	0.0753	-0.0519	1.3420	0.0706	0.0055	-0.0088	0	0.0121
2	0.6651	0.0970	0.0126	0.0753	-0.0519	1.3420	0.0706	0	-0.0061	0	0.0121
3	0.6530	0.1031	0.0126	0.0753	-0.0519	1.3420	0.0706	0.0055	0	-0.0088	0.0121
4	0.6530	0.1031	0.0126	0.0753	-0.0519	1.3420	0.0706	0	0	-0.0061	0.0121
5	0.6544	0.1008	0.0178	0.0798	-0.0550	1.3450	0.0751	-0.0104	0	0	0.0104

Table 9.4. Models from PHREEQC calculations for well Navajo National Monument water. Units are millimoles per liter. Results from the hand/spreadsheet calculation are listed for comparison.

areas, possibly contribute to the pollution. Extensive monitoring activities and research have been conducted by the US Geological Survey, US EPA, and industry.

There are many practical reasons why migration of the Pinal Creek acid plume and associated metals needs to be predicted from models:

1. the responsible parties need to report to EPA and state regulators quarterly on the condition of the pollution;

2. models are needed to predict and evaluate the effectiveness of remedial pumping programs at the site;

3. because the contamination has occurred over a long time and may have come from many different sources, lawsuits have ensued to proportion costs among the mining companies involved.

As we have shown in Chapters 6 and 7, the migration of an acid plume from acid mine drainage is a very complex process. However, only K_d-based models have been used to simulate the acid plume and metal transport at this site for regulatory reporting. See Chapter 10 for a criticism of K_d-based transport models. As we contend in this book, accurate model simulations are only possible if we understand what geochemical reactions have occurred and are occurring in the aquifer.

Glynn and Brown (1996) used inverse mass balance modeling to help identify (1) the mass transfer reactions and (2) the extent of the mass transfer reactions at the Pinal Creek site. They chose a flow path from well 402 to well 503, which are about 5.6 km apart. Well 402 has a pH of 4.13 and is high in Fe and Al. Well 503 has a pH of 5.59 and the Fe and Al are drastically lower (Table 9.5). Hence, this flow path may indicate the mass transfer reactions when acidic groundwater is neutralized by reactions with aquifer materials. Because they suspect that dilution also occurs along this flow path, a background well 504 is also included in calculating the mixing ratios. The inverse problem is therefore essentially

$$\text{well 402} + \text{well 504} + \text{reactants} \rightarrow \text{well 503} + \text{products} \qquad (9.2)$$

Note that this is a situation where the solute concentrations vary in both space *and* time. Therefore, the chemical steady state assumption that is necessary for inverse modeling may be violated.

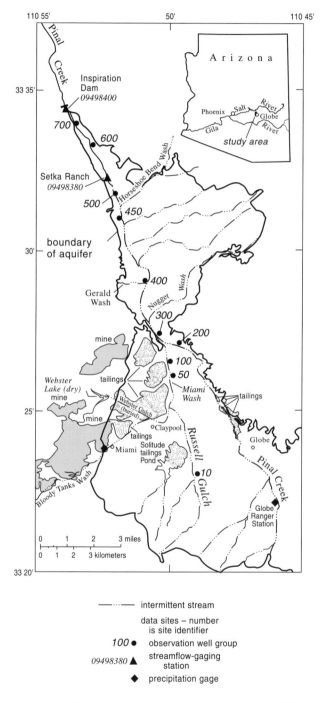

Figure 9.2. The Pinal Creek site, after Brown *et al.* (1998).

Component	Well 402	Well 503	Well 504
pH	4.13	5.59	7.05
S	3260	2350	14.2
SiO_2	85.6	91.8	27
Ca	502	634	44.6
Na	121	86	19.8
K	7 est.	5 est.	2.1
DO	0.3	<0.1	6.64
Mg	161	200	15.6
Fe	591	<0.1	0.004
Al	18.4	2.3 est.	<0.01
Mn	71.6	116	<0.0001
Cl	140	112	9.7
Alkalinity	50	66	227

Table 9.5. Chemical compositions of groundwater at Pinal Creek, Arizona. Concentrations are in milligrams per liter. Groundwater flows from Wells 402 to 503.

Glynn and Brown (1996) used 11 mass balance constraints on Cl, Ca, Mg, Na, Al, Si, Fe, Mn, C, redox state, and S. Fourteen phases were considered in the model, calcite, goethite, gypsum, SiO_2, dolomite, MnO_2, $MnCO_3$, anorthite, gibbsite, $Mn(OH)_3$, tremolite, $AlOHSO_4$, biotite, and K-montmorillonite. Note that all elements in the constraints must be contained in at least one phase; otherwise the sink and source can only be from the mixing fluids. In this case, none of the phases contain Cl or Na. The net changes of Cl or Na then determine the dilution from background groundwater.

PHREEQC allows us to specify the direction of mass transfer reactions. The "−" sign in Table 9.6 designates precipitation only and "+" dissolution only. (NETPATH also allows us to "force" a phase to be included in every model.) A ±5% analytical uncertainty is assigned to all analytes except for K, for which an uncertainty of ±25% was given.

Seven tentative inverse mass balance models were produced by PHREEQC with the mass balance and phase constraints given (Table 9.6). A mixing fraction of 0.258 for well 504 water (hence 0.742 for well 403 water) was determined based on the conservative chemical Cl. Among the seven models, most are essentially combinations of proportions of different minerals. With different input constraints, more inverse models can be produced, which also produce the compositional differences between the initial and final wells.

By now, readers may realize that the results from Glynn and Brown (1996) are fairly inconclusive. As emphasized above, the modeling output depends on the mineral and gas phases assumed at the outset in the input. Possible models can be considerably narrowed down if detailed mineralogical studies are conducted.

Phase	Model 1	Model 2	Model 3	Model 4	Model 5	Model 6	Model 7
504 mf	0.258	0.258	0.258	0.258	0.258	0.258	0.258
Dolomite +	0.398		2.191		0.29		2.291
Gypsum −							
Goethite −	−8.292	−8.292	−8.292	−8.291	−8.292	−8.292	−9.074
Calcite +	3.588	4.383		1.626	3.817	3.991	
Anorthite +			2.512	2.202			2.691
AlOHSO$_4$ −					−0.423	−0.423	
Gibbsite	−0.423	−0.423	−5.448	−4.827			−2.167
SiO$_2$	−3.649	−3.649	−5.803	−8.687	−3.165	−3.629	
MnCO$_3$	−2.757	−2.757	−2.756		−2.769	−2.364	−3.253
MnO$_2$ +	3.903	3.903	3.903	6.659	3.903	4.309	4.294
Mn(OH)$_3$ −				−5.512		−0.811	
Tremolite +	0.500	0.500	0.142	0.580	0.440	0.498	
Biotite +							0.521
K-mont −							−1.796
Alk. change	0.214	0.213	0.213	0.213	0.213	0.213	0.190
H+ consum.	−0.279	−0.281	−0.272	−0.266	−0.271	−0.270	−0.298

Table 9.6. Results of inverse modeling at Pinal Creek, Arizona. Concentrations are millimoles per liter.

9.5 ^{14}C dating, Black Mesa, Arizona

An important application of inverse mass balance modeling is to ^{14}C age correction. ^{14}C is a useful environmental tracer to study groundwater systems. With a half-life of about 5 730 years (Goodwin, 1962), it can give dates from a few thousands to about 45 000 years. Large corrections to apparent ^{14}C ages are often required because dissolved inorganic carbon (DIC) in groundwater samples is a mixture of carbon from different sources. To calculate the proportions of DIC from different sources, various forms of inverse modeling of groundwater genesis is employed. Unfortunately, most applications use one-size-fits-all mass transfer models that do not consider the geology of the groundwater aquifer of interest. However, this is the type of application for which inverse mass balance modeling is well suited (Zhu and Murphy, 2000).

There is an extensive literature on applying inverse mass balance modeling to ^{14}C dating of groundwater, mostly by Plummer and co-workers (Plummer *et al.*, 1990, 1991, 1994; Wigley *et al.*, 1978). The code NETPATH has been extensively programmed to do age corrections. Many popular off-the-shelf correction models are built into NETPATH. Again, readers are cautioned that it is the modeler, not the modeling code, that produces a model (see Chapter 4). One can use NETPATH to devise any correction models that are consistent with local hydrogeology and geochemistry. ^{14}C dating of groundwater is a very complex problem. The readers are again referred to read the study of the Madison aquifer by Plummer *et al.* (1990) as a guide. Here, we give a simpler example from Zhu (2000).

This example follows §9.3, where we used spreadsheet and NETPATH calculations to deduce the mass transfer reactions along the evolutionary path from snow melt to water in the saturated zone of recharge area, specifically in the Navajo National Monument well. Now, we want to calculate the travel time from Navajo National Monument to well

Component	Final	Initial
C	1.3603	2.1187
S	0.0302	0.0284
Si	0.3329	0.0932
Ca	0.0923	0.8485
Na	1.4356	0.1510
K	0.0154	0.0274

Table 9.7. Chemical compositions of initial and final wells. Concentrations are in millimoles per liter.

Phases	Model 1	Model 2
calcite	-0.75839	-0.75839
gypsum	0.00177	0.00177
kaolinite	-0.06554	
Illite	-0.02003	-0.02003
Na–Ca exch	-0.35353	-0.35353
An$_{38}$	0.93164	0.93164
SiO$_2$		-2.13108

Table 9.8. NETPATH modeling results. Positive numbers denote quantities (millimoles per liter) dissolved and negative numbers denote millimoles precipitated.

NAV 9, located about 15 km south from Navajo National Monument well (Figure 9.1). Following Plummer *et al.* (1991), we call the first well the "initial" well and the second the "final" well.

The Navajo sandstone in the Black Mesa area is an aeolian quartz sandstone (Harshbarger *et al.*, 1957). Besides quartz, other types of sand grains include mostly plagioclase and, to a lesser extent, orthoclase (Dulaney, 1989; Harshbarger *et al.*, 1957). The sandstone is cemented by calcite and, to a lesser extent, by silica, and red iron oxide rims are common on the sand grains.

Numerical simulations of flow, particle tracking, and ^{14}C transport show that the two samples are approximately along the same flow path, and for this regional aquifer with relatively homogeneous lithology, diffusion, and dispersion are negligible for long-term transport.

Six components, Na, Ca, K, C, S, and Si, and seven possible phases, calcite, gypsum, kaolinite, illite, Na–Ca exchange, and plagioclase with An$_{38}$, were used to construct the reactions that might have occurred along this flow path (Table 9.7). NETPATH produced two models (Table 9.8). Model 1 shows precipitation of calcite, kaolinite, illite, and dissolution of plagioclase and gypsum as well as Ca for Na ion-exchange. Model 2 precipitates quartz rather than kaolinite.

Zhu (2000) used the modeling results to calculate travel time. The initial well has a radiocarbon activity of 72.9% modern, and final well 7%. Precipitation of calcite results in isotopic fractionation, which NETPATH treats as a Rayleigh distillation process (Plummer *et al.*, 1991; Wigley *et al.*, 1978). After the mass balance calculations, there

---

Chemical non-steady state for chloride

Chloride is usually a major constituent in groundwater and is widely considered a conservative tracer. In the N aquifer, Cl^- concentrations are considerably higher in Holocene water than in late Pleistocene water (Figure 9.3). The groundwater ages indicated on the horizontal axis are results from inverse mass balance modeling and age corrections (Zhu, 2000). These age data are also supported by the δD and $\delta^{18}O$ data of the same samples. It is generally known that Pleistocene water has depleted H and O stable isotope values with respect to recent water because of a cooler and more humid climate in the late Pleistocene (Merlivat and Jouzel, 1979).

It should be noted that processes that could possibly affect Cl^- concentrations in the aquifer, for example, (1) diffusion or leakage from the overlying D aquifer, (2) dissolution of Cl^- bearing minerals (e.g., biotite) or fluid inclusions, or (3) sampling artifacts such as partial screening in the D or overlying Carmel Formation, all tend to increase Cl^- concentration in the downgradient and older groundwater.

Therefore, the large variation in Cl^-, particularly the decrease along the flow path, is likely to be a result of paleoclimate changes during late Pleistocene and Holocene. The temperature in the late Pleistocene in the area was 5 to 6 degrees cooler during the last glacial maximum than today (Zhu *et al.*, 1998, and references therein). There was possibly a shift from summer precipitation to the winter. As a result, there was less evaporation and higher recharge than today. Additionally, the atmospheric deposition rate of Cl^- may also be different in the late Pleistocene. All these could contribute to a notably lower Cl^- in older groundwater. This is a case in which Cl^- is not in chemical steady state. Hence, Cl^- was not used in the inverse mass balance model.

---

are still differences between the observed $\delta^{13}C$ and computed $\delta^{13}C$. These differences are reconciled by adjusting the amount of carbon engaged in carbon isotopic exchange between aqueous carbonate species and calcite (1.16 mmol L^{-1}). Plummer *et al.* (1991) regard isotopic exchange as precipitation and dissolution of equal amounts of carbonate (recrystallization). The inverse modeling gives a travel time of 13 744 years while the apparent decay age is 19 370 years.

Note that the carbon isotopes are solely constrained by the DIC and $\delta^{13}C$ values. Therefore, the two inverse mass balance models give the same results with respect to calculated travel time. Precipitation of illite, quartz, or kaolinite, the dissolution of gypsum and plagioclase, and Ca–Na exchange are needed to interpret the increases of Na^+, SO_4^{2-}, and $SiO_2(aq)$, and decreases of Ca^{2+} and K^+ in downgradient wells.

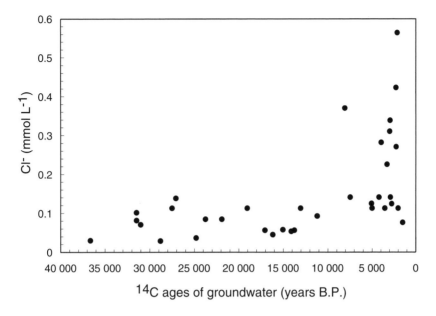

Figure 9.3. Variation of groundwater Cl^- concentrations at Black Mesa with time.

9.6   Estimate of Microbial Metabolism Rates in Deep Aquifers

Microbial-mediated reactions in the subsurface could have a significant influence on the groundwater geochemistry (Chapelle, 2000). However, estimates of the *in situ* microbial metabolism rates in deep aquifers are difficult. Laboratory studies may over-estimate the *in situ* rates (Murphy and Schramke, 1998). Additionally, although various techniques and precautions are used during drilling, sampling, and laboratory handling, contamination-free samples are not assured. Studies of geochemical indicators of microbial activities provide an important part in understanding microbial activities in deep aquifers. In particular, geochemical modeling provides an estimate of *in situ* rates of microbial metabolism. Here, we give two examples from the literature that used the inverse modeling tool to estimate microbial activities in deep aquifers in the Coastal Plain of South Carolina.

9.6.1   Chapelle and Lovley (1990)

Chapelle and Lovley (1990) studied anaerobic aquifers of the Atlantic Coastal Plain in South Carolina. The aquifers include Black Creek, Middendorf, and Cape Fear Formations. Groundwater in these aquifers is depleted in dissolved oxygen. Microbial metabolism consumes the organic matter in the aquifer, produces CO_2, and reduces ferric iron or sulfate, which are terminal electron acceptors.

Observed geochemical changes along flow paths are:

- increase in total dissolved inorganic carbon (DIC);

- increase of Na^+, while Ca^{2+} and Mg^{2+} concentrations remain the same;

- pH remains approximately the same.

From these trends, Chapelle and Lovley (1990) inferred the following mass transfer reactions.

1. Microbially produced CO_2 results in dissolution of carbonate shell,

$$CO_2 + Ca_{.98}Mg_{.02}CO_3 + H_2O \rightarrow 0.98Ca^{2+} + 0.02Mg^{2+} + 2HCO_3^- \quad (9.3)$$

The overall change of DIC (ΔM_C) is the sum of microbially produced CO_2 and dissolution of shell materials (m_{sm}):

$$\Delta M_C = m_{CO_2} + m_{sm} \quad (9.4)$$

where M denotes moles per liter of the chemical in the subscript.

2. Cation-exchange reactions

$$Ca^{2+} + Na_2(clay) \rightarrow Ca(clay) + 2Na^+ \quad (9.5)$$

$$Mg^{2+} + Na_2(clay) \rightarrow Mg(clay) + 2Na^+ \quad (9.6)$$

and the net mass transfer along the flow path for Ca and Mg are

$$\Delta M_{Ca} = 0.98m_{sm} - m_{Ca(clay)} \quad (9.7)$$

$$\Delta M_{Mg} = 0.02m_{sm} - m_{Mg(clay)} \quad (9.8)$$

3. The change of chloride concentrations along the flow path is thought to be due to mixing with pore fluids of incomplete flushing of seawater in the marine sediments

$$\Delta M_{Cl} = m_{sw} \quad (9.9)$$

and for sodium the mass transfer is

$$\Delta M_{Na} = m_{sw} + 2m_{Ca(Clay)} + 2m_{Mg(clay)} \quad (9.10)$$

Combining all the equations above, the overall CO_2 production from microbial activities can be calculated from:

$$m_{CO_2} = \Delta M_C - \Delta M_{Ca} - \Delta M_{Mg} - (\Delta M_{Na} - \Delta M_{Cl})/2 \qquad (9.11)$$

The actual calculations were performed using BALANCE, an earlier inverse mass balance modeling code that is now superseded by NETPATH and PHREEQC. Chapelle and Lovley (1990) then calculated the time interval in these segments from flow velocities calculated by a numerical groundwater flow model and the length of the flow path. The total CO_2 production rate from oxidation of organic matter is

$$\text{metabolism rate} = m_{CO_2}/\Delta t \qquad (9.12)$$

The calculated rates are 10^{-4} to 10^{-6} millimoles CO_2 per liter per year. These *in situ* rates are two to four orders of magnitude slower than rates measured from laboratory incubation. However, the *in situ* rates are more consistent with the sediment ages of about 80 ma. The discrepancy between *in situ* and laboratory rates suggests that the latter may greatly overestimate the *in situ* microbial metabolism rate in deep aquifers.

9.6.2   Murphy and Schramke (1998)

In a more recent study, Murphy and Schramke (1998) estimated microbial respiration rates from geochemical modeling for individual electron acceptors O_2, MnO_2, Fe^{3+}, and SO_4^{2-}. They were able to do so because they used additional constraints, stable isotopes ^{13}C and ^{34}S.

The following reactions were used for the respiration of individual electron acceptors:
-1pt

$$CH_2O + O_2 \rightarrow H_2CO_3 \qquad (9.13)$$

$$CH_2O + \tfrac{1}{2}SO_4^{2-} + H^+ \rightarrow \tfrac{1}{2}H_2S + H_2CO_3 \qquad (9.14)$$

$$CH_2O + 4Fe^{3+} + 2H_2O \rightarrow 4Fe^{2+} + H_2CO_3 + 4H^+ \qquad (9.15)$$

$$CH_2O + 2MnO_2 + 4H^+ \rightarrow 2Mn^{2+} + H_2CO_3 + 2H_2O \qquad (9.16)$$

Murphy and Schramke (1998) first used the $\delta^{13}C$ values to show that the increase of HCO_3^- is most likely by microbially mediated oxidation of organic carbon in lignite, not from dissolution of calcite or diffusion of CO_2 from neighboring aquifers. Therefore, the total change in dissolved inorganic carbon (DIC) can be calculated from the equation

$$\Delta C_T = C_{O_2} + 2C_{SO_4^{2-}} + \tfrac{1}{4}C_{Fe^{3+}} + \tfrac{1}{2}C_{MnO_2} \qquad (9.17)$$

where C_T denotes the concentration of DIC. The terms on the right hand side correspond to the concentration of DIC that is produced by microbial respiration using O_2, SO_4^{2-}, Fe^{3+} and MnO_2, respectively. The individual respiration rates may be calculated if each term on the right hand side of the equation can be estimated.

The amounts of CO_2 produced by respiration of oxygen are monitored from the depletion of dissolved oxygen along the flow path. The amount of CO_2 produced by respiration of Mn^{4+} is calculated from the dissolution of MnO_2, which in turn is constrained by the increase of Mn^{2+} along the flow path.

The amounts of organic carbon oxidized by Fe^{3+} and SO_4^{2-} are more difficult to estimate. The total precipitated FeS is calculated by NETPATH as described below. However, Murphy and Schramke (1998) believe that two reaction pathways are possible for FeS precipitation. In Pathway 1, FeS is produced by oxidation of organic sulfur to sulfate and subsequently reduced to HS^- by bacteria. In Pathway 2, FeS is formed by direct reaction of Fe^{2+} with pyritic sulfur in the organic sulfur phase. The two pathways should differ in ^{34}S values because isotopic fractionation is required in Pathway 1. NETPATH was used to model the isotopic evolution for the two reaction pathways, but the modeling results for neither pathway compared well with the measured groundwater $\delta^{34}S$ values. Murphy and Schramke concluded that a combination of the two reaction pathways is needed for interpretation of the measured groundwater isotopes.

By constraining the fraction of the total FeS that is from SO_4^{2-} reduction based on isotope mass balance, the amount of CO_2 produced from respiration of Fe^{3+} can then be calculated from the remaining portion of the total organic carbon oxidized.

For the general chemical mass balance, Murphy and Schramke (1998) included:

- exchange of Ca and Mg for Na;

- dissolution of K- and Ca-feldspars;

- oxidation of an organic S phase;

- oxidation of an organic C phase;

- dissolution of quartz;

- dissolution of gibbsite;

- dissolution of MnO_2;

- dissolution of $Fe(OH)_3$;

- precipitation of FeS;

- precipitation of Ca-nontronite (an Fe-containing phase).

Respiration rates are calculated by dividing each term in Equation (9.17) by the ^{14}C ages. The calculated respiration rates for terminal electron acceptors O_2, MnO_2, Fe^{3+}, and SO_4^{2-} are 10^{-6}, 10^{-8}, 10^{-4}, and 10^{-6} per liter per year, respectively. These rates are again much lower than those estimated from biological methods such as adding substrate or electron acceptors, and are lower than the near-surface sediments.

10

Coupled Reactive Transport Models

10.1 Introduction

The fate and transport of contaminants in subsurface environments are known to be controlled by complex physical, chemical, and biological processes that are coupled in nature. However, most models described in the preceding chapters only simulate geochemical reactions in a static system, resembling the contents of a beaker. Hydraulic effects, such as advection and dispersion, are not taken into account. In reality, the transport processes bring about mass fluxes between spatial domains, which disturb the equilibrium or chemical steady state and result in chemical reactions. Therefore, the direction and extent of chemical reactions can be determined by transport processes in addition to thermodynamics and chemical kinetics. As discussed before, the regulatory environment requires predictions of concentration distribution in space and time. Because static geochemical models are unable to deal with chemical reactions on temporal and spatial coordinates, they are of little direct use to the end results that are sought.

A natural response to the limitations of both geochemical equilibrium models and the solute transport models (see §10.3 for a discussion) is to couple the two. Over the last two decades, a number of models that couple advective–dispersive–diffusive transport with fully speciated chemical reactions have been developed (see reviews by Engesgaard and Christensen, 1988; Grove and Stollenwerk, 1987; Mangold and Tsang, 1991). In the coupled models, the solute transport and chemical equilibrium equations are simultaneously evaluated.

These models, however, are mostly at the research and development stage, and most model applications are "example" calculations demonstrating code capability. Applications as a part of the fulfilment of regulatory requirements are just beginning to appear at the time of writing. Coupled models are complex and computationally expensive. A coupled model requires the modeler to have a good knowledge of both hydrogeology and geochemistry. Additionally, the data requirements, both field data and thermodynamic and kinetic properties, are formidable.

Theories of coupled reactive transport models can be found in Cederberg *et al.* (1985), Lichtner (1996), Raffensperger and Garven (1995), Steefel and MacQuarrie (1996), Walsh (1983), Walter *et al.* (1994), and Yeh and Tripathi (1989). Here, we only give a rudimentary introduction to the concept.

10.2   Multi-component Reactive Transport Models

Consider advective–dispersive–reactive (ADR) transport in one-dimensional saturated porous media, the transport equation is (Bear, 1972),

$$\frac{\partial C_i}{\partial t} = D_l \frac{\partial^2 C_i}{\partial x^2} - v \frac{\partial C_i}{\partial x} + \sum_{k=1}^{n} R_i \qquad (10.1)$$

where C_i represents the concentrations of component i in groundwater (M L^{-3}). t denotes time (T), and x distance (L). v stands for the average linear velocity of ground-water (L T^{-1}) and D_l for the longitudinal hydrodynamic dispersion coefficient (L^2 T^{-1}) (Bear, 1972). R_i denotes the rate of addition or removal of i to or from groundwater due to reaction k, and n represents the total number of reactions affecting i.

Equation (10.1) is a statement of mass balance on a per volume of groundwater basis for a given component i. The first two terms on the right hand side of the equation refer to the net influxes of mass due to dispersive and advective transport in groundwater.

The reaction term R_i describes the loss or gain of component i in groundwater due to the mass transfer between groundwater and immobile solid phases. From the transport point of view, mass transfer reactions between aquifer matrix and groundwater represent possible sinks and sources for the contaminants so that this term is sometimes called the source/sink term.

In the coupled models, the reaction term is solved separately for each cell by using a chemical module (e.g., MINTEQA2, PHREEQE), in which the partitioning of chemical components between the solid phases and aqueous solutions are calculated based on aqueous speciation, solubility, and surface complexation reactions. These chemical reactions are solved through the mass-balance and mass-action equations. A common strategy to solve the ADR equation is various forms of split-operator schemes (Steefel and MacQuarrie, 1996). For example, in PHREEQC, within each time step, the advec-tive transport is calculated first by using an upwind finite difference scheme. This is followed immediately by a calculation of chemical reactions. Then, dispersive trans-port is calculated using a central difference scheme (or mixing cell method; see Appelo and Postma, 1993). This is again followed by a calculation of chemical reactions. In some codes, iterations between the physical and chemical steps are used (e.g., Yeh and Tripathi, 1989).

As discussed in Chapter 2, we reserve the term "coupled transport model" to multiple component-multiple species reactive mass transport models such as PHREEQC described above. In coupled models, two set of equations are solved together through some coupling schemes. In the case of coupled reactive transport models, two sets of mathematical equations – the partial differential equations that describe advective–dispersive transport and the algebraic equations that describe chemical reactions – are solved.

Additional sets of equations that describe additional processes can be coupled and solved together. For example, the conservation of thermal energy and groundwater flow equations are coupled to the ADR equation and chemical mass balance equations in the work of Raffensperger (1995).

10.3 Isotherm-based Reactive Transport Models

Nearly all "reactive transport models" used in the regulatory environment, however, are based on empirical isotherms. The models take into account the effects of hydrologic processes due to the movement of groundwater such as advection and dispersion, but the effects of chemical reactions are typically described by an isotherm linking the concentrations of i in solid mass to that in groundwater,

$$S_i = f(C_i) \tag{10.2}$$

where S_i and C_i represent the concentrations of component i bound to the solid matrix $(M\ M^{-1})$ and in groundwater $(M\ L^{-3})$, respectively.

The isotherms are incorporated into the transport models through the retardation factor R. The one-dimensional ADR equation for saturated media becomes

$$R_i \frac{\partial C_i}{\partial t} = D_l \frac{\partial^2 C_i}{\partial x^2} - v \frac{\partial C_i}{\partial x} \tag{10.3}$$

where R is defined as (Fetter, 1999)

$$R_i = 1 + \frac{\rho_b}{\theta} \frac{\partial S_i}{\partial C_i} \tag{10.4}$$

ρ_b denotes the bulk density of the aquifer $(M\ L^{-3})$ and θ the effective porosity (dimensionless). The ADR equation thus becomes more mathematically solvable.

It is important to note that Equations (10.2) and (10.3) are based on the local equilibrium condition which assumes that reaction rates are fast in relation to groundwater flow rates (Cherry *et al.*, 1984). Other assumptions in Equation (10.2) include the reversibility of reaction and the absence of competing species for the same surface sites.

To facilitate the discussion below, we need, once again, to define the terminology first. By our definition, the isotherm-based reactive transport models are not "coupled models" because only one set of equations, namely partial differential equations for transport, is solved. The chemical processes are simulated according to empirical parameters rather than according to thermodynamics and chemical kinetics.

10.3.1 Linear Isotherm, K_d

The distribution coefficient or K_d is described in §7.2.3. K_d is factored into the transport equation through a *retardation factor*, defined as

$$R_i = 1 + \frac{\rho_b}{\theta} K_d \tag{10.5}$$

10.3.2   Freundlich Isotherm

The retardation factor consistent with the Freundlich isotherm becomes

$$R_i = 1 + \frac{\rho_b}{\theta} K_F n C_i^{n-1} \tag{10.6}$$

In this case, the retardation factor is a function of solute concentrations. It will increase at low concentrations. Columns with pulse inputs will have breakthrough curves with long tails (Grove and Stollenwerk, 1987).

10.3.3   Langmuir Isotherm

The retardation factor consistent with the Langmuir isotherm is derived as

$$R_i = 1 + \frac{\rho_b}{\theta} \left[\frac{K_L b}{(1 + K_L C_i)^2} \right] \tag{10.7}$$

The retardation factor, R_i, represents the impediment of a concentration front with respect to the bulk mass of water,

$$R_i = \frac{v}{v_i} \tag{10.8}$$

where v_i is the velocity of relative concentration C/C_o value of 0.5 of the constituent i when dispersion is considered. The reciprocal of the R_i ($= v_i/v$) is known as the relative velocity. R_i can also be viewed as the shortened distance that a solute has traveled at a given time or the reduction of values of transport parameter D and v (Bear, 1972; Domenico and Schwartz, 1998).

10.3.4   Applicability of the Isotherm or Retardation-factor-based Reactive Transport Models

Although isotherms are a useful descriptions of the partitioning of solute between aqueous solution and solids, and the retardation-factor-based equation (10.3) made the ADR equation mathematically solvable, the applicability of these "reactive transport models" to environmental problems is quite limited (Bethke and Brady, 2000; Cederberg *et al.*, 1985; Reardon, 1981; Zhu and Burden, 2001; Zhu *et al.*, 2001a). These models are too simplistic regarding the assumptions of the geochemical reactions at many sites, perhaps with only a few exceptions (such as the adsorption of hydrophobic contaminants onto organic or organically coated substances; Stumm and Morgan, 1996).

Even though the limitations of isotherm-based transport models are well known to academics, and detailed from time to time in the literature (see the references cited above), isotherm-based reactive transport models nevertheless remain the mainstay in everyday environmental practice. Therefore, we feel the need to elucidate the weakness and limitations again.

Adsorption Isotherm or Sorption Isotherms?

There appears to be a disparity in terminology of *isotherms* between the hydrogeology and chemistry literature. While chemistry textbooks describe *adsorption* isotherms (e.g., Drever, 1988, p. 362; Stumm and Morgan, 1996, p. 521), the hydrogeology literature uses the term *sorption* isotherms (e.g., Domenico and Schwartz, 1998, p.299; Fetter, 1999, p.122; Freeze and Cherry, 1979, p.403). Manuals for popular transport codes also describe "sorption isotherms" (Zheng, 1990, pp.2–9). According to Sposito (1984), the term "adsorption" includes all reactions on the two-dimensional solid–water interface: physical and chemical adsorption and ion-exchange; the term "sorption" refers to all processes that transfer an ion from aqueous to solid phase.

If precipitation–dissolution reactions dominate (this is sorption by Sposito's definition), there are simply no differentiable isotherms (defined as a functional relationship between S and C) relating solid and aqueous concentrations (Bryant *et al.*, 1987). The above-mentioned retardation factor formulations (equation (10.3)) are not applicable. A more complete form of the ADR equation may be (Fetter, 1993, p.115),

$$\frac{\partial C_i}{\partial t} = D_l \frac{\partial^2 C_i}{\partial x^2} - v \frac{\partial C_i}{\partial x} + \frac{\rho_b}{\theta} \frac{\partial S_i}{\partial t} + \left(\frac{\partial C_i}{\partial t} \right)_{\text{rxn}} \tag{10.9}$$

where the rightmost term takes into account the effects of non-adsorption chemical reactions.

Cherry *et al.* (1984) noted that:

> ... therefore, the use of predictive methods that represent adsorption with exclusion of the effects of other processes or the other processes with the exclusion of adsorption is usually based on the convenience of conceptualization and computation rather than a quest for realism.

Comparing the Langmuir Isotherm with the Surface Complexation Model

The simplifications and assumptions of the retardation factor approach in treating chemical reactions can be illustrated by comparing the one-site Langmuir isotherm and surface complexation formulation for adsorption of a metal onto iron oxide with one type of surface sites.

Farley *et al.* (1985) show that the adsorption of a metal onto an oxide surface can be formulated according to the surface complexation model,

$$S_i' = \frac{S_T \frac{K^{app}}{[H^+]\gamma_{H^+}} C_i \gamma_i}{1 + \frac{K^{app}}{[H^+]\gamma_{H^+}} C_i \gamma_i} \tag{10.10}$$

where S_i' stands for the concentration of surface species containing i in moles per liter. Note that S_i was defined before and has a unit of $(M \, M^{-1})$. To convert to pore fluid volume based units $(M \, L^{-3})$, we need to do the following calculation. In a porous medium with porosity of θ and dry bulk density of ρ_b,

$$S_i' = S_i \frac{\rho_b}{\theta} \frac{1}{\text{atomic weight of } i \times 1000} \tag{10.11}$$

S_T denotes the total surface sites per liter of fluid in the porous media in moles per liter. C_i here stands for the concentration of the *"free"* ion; $[H^+]$ is the concentration of the H^+ ion. K^{app} is the *apparent* equilibrium constant for surface complexation,

$$K^{app} = K^{intr} K^{Coul} \qquad (10.12)$$

where K^{intr} is the *intrinsic* equilibrium constant independent of the solution chemistry and surface charges and K^{Coul} represents the electrostatic interaction at the solid surface (see Chapter 7 for more details).

γ_i stands for activity coefficients defined by

$$a_i = C_i \gamma_i \qquad (10.13)$$

Note that Equation (10.10) has the form of a Langmuir isotherm if

$$\alpha = \frac{K^{app} \gamma_i}{[H^+] \gamma_{H^+}} \qquad (10.14)$$

is constant and equivalent to K_L.

It is now apparent from Equation (10.14) that there are numerous assumptions in a Langmuir isotherm and associated transport models

- Constant solution *p*H. A constant and uniform Langmuir isotherm means that *p*H is the same for all simulated domains and does not change with time.

- Solution chemistry. Note that C_i in the surface complexation model formulation is the *free* ion concentration, while in the Langmuir isotherm it is the total analytical concentration. In reality, the formation of aqueous complexes will change the surface adsorption (affecting the term C_i). The speciation of i may also change with time.

- Constant ionic strength (affecting the K^{Coul} term and also activity coefficient for i and H^+).

Additionally, S_T is an aquifer specific property that is related to the amount, species, and surface properties of iron oxides in the aquifer. Hence, the influence of spatial chemical heterogeneity is apparent. When iron oxide compositions in the samples, even from the same site, are different, different isotherms will result. Therefore, isotherms that does not specify these conditions can vary significantly.

Groundwater systems that satisfy these chemical requirements are not common. Transport models that incorporate the Langmuir isotherms are likely to provide predictions of contaminant behaviors that may not be realistic.

Large Variability of Parameters

Because isotherms are empirical, a parameter used for one field site cannot generally be used for another – even for materials from the same site. K_d values, for example, measured for radionuclide partitioning between volcanic tuff from the Yucca Mountain area, southern Nevada and well J-13 water, vary by orders of magnitude (Thomas, 1987).

The limitations of K_d rest on the fact that these are phenomenological parameters that does not take into account the many factors that affect sorption (Cherry *et al.*, 1984; Davis and Kent, 1990).

Unlimited Sources and Sinks

As often noted, an unlimited amount of solutes are allowed to sorb onto solids in the linear and Freudlich isotherm formulations, whereas in reality only limited sites are available in a given system.

Interactions between Solutes

Another drawback of most conventional reactive solute transport model is that it only evaluates one solute at a time. To emphasize this, we deliberately used the subscript i in the above equations. The interactions between solutes, i.e., reactions of two solutes to form a solid precipitate, competitive sorption of metals onto mineral surfaces, or co-precipitation, cannot be evaluated. Although some isotherm-based transport codes have included the transport of several components, the chemical modules are nevertheless too simple to allow them to interact.

Temporal Variability

The distribution coefficients also change with time and space due to chemical reaction progress. Although transport codes can be modified to assign spatially variable K_d values, the variation with time cannot be set *a priori*.

Even though the pitfalls of the isotherm-based models are well known, it is still widely used. However, it has been shown, case by case, that the modeling results differ greatly from experience years later (Bethke and Brady, 2000; Kohler *et al.*, 1996; National Research Council, 1994). Typically, overly optimistic results are predicted which has led to an underestimate of remediation costs and clean-up time. They are continually used because of

- mathematical simplicity;

- the assurance of an answer in a relatively short period of time, and hence low cost;

- easy measurements of K_d values from batch and column tests; and

- a practice currently accepted by regulatory bodies.

10.4   A Simple Example

To illustrate the concept of coupled reactive transport models and differentiate them from static geochemical and solute transport models, we use a simple example modified from

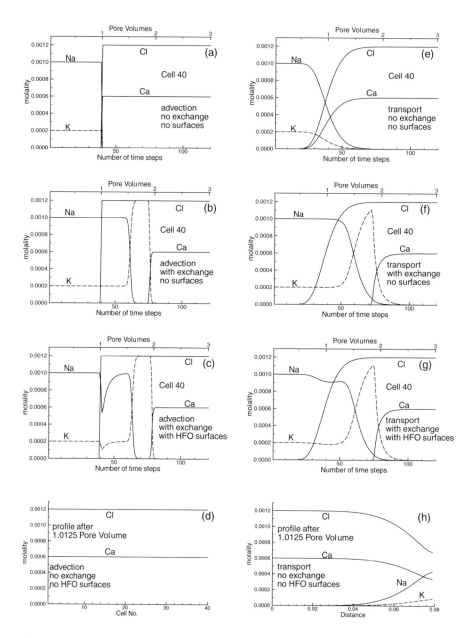

Figure 10.1. An example of the reactive transport capabilities of PHREEQC. Pure advection (plug flow; ADVECTION option) are shown in (a)–(d); advection plus dispersion and diffusion (TRANSPORT option) are shown in (e)–(h).

```
PRINT
        status false
TITLE Example 11.--Transport and ion exchange.
SOLUTION 0   CaCl2
        units           mmol/kgw
        temp            25.0
        pH              7.0       charge
        pe              12.5      O2(g)    -0.68
        Ca              0.6
        Cl              1.2
SOLUTION 1-40   Initial solution for column
        units           mmol/kgw
        temp            25.0
        pH              7.0       charge
        pe              12.5      O2(g)    -0.68
        Na              1.0
        K               0.2
        N(5)            1.2
#EQUILIBRIUM_PHASES        1
#        Fe(OH)3(a)       0.0
EXCHANGE 1-40
        equilibrate 1
        X               0.0011
#SURFACE    1
#        -equilibrate with solution 1-40
#        Hfo_wOH Fe(OH)3(a) equilibrium_phase  0.2  53400
ADVECTION
        -cells          40
        -shifts         120
#       -punch_cells    1-40
#       -punch_frequency 60
        -print_cells    40
        -print_frequency 20
SELECTED_OUTPUT
        -file           ex11adv.sel
        -reset          false
        -step
#       -distance
        -totals         Na Cl K Ca
USER_PUNCH
        -heading  Pore_vol
        10 PUNCH (STEP_NO + .5) / 40.
END
SOLUTION 1-40   Initial solution for column
        units           mmol/kgw
        temp            25.0
        pH              7.0       charge
        pe              12.5      O2(g)    -0.68
        Na              1.0
        K               0.2
        N(5)            1.2
```

Table 10.1. A modified PHREEQC input file for Example 11. (Continued on next page.)

```
#EQUILIBRIUM_PHASES          1
#        Fe(OH)3(a)          0.0
EXCHANGE 1-40
         equilibrate 1
         X                   0.0011
#SURFACE  1
#        -equilibrate with solution 1-40
#        Hfo_wOH Fe(OH)3(a) equilibrium_phase  0.2   53400
TRANSPORT
         -cells             40
         -length            0.002
         -shifts            120
         -time_step         720.0
         -flow_direction    forward
         -boundary_cond     flux        flux
         -diffc             0.0e-9
         -dispersivity      0.002
         -correct_disp      true
#        -punch             1-40
#        -punch_frequency 60
         -print             40
         -print_frequency 20
SELECTED_OUTPUT
         -file              ex11trn.sel
         -reset             false
         -step
#        -distance
         -totals            Na Cl K Ca
END
```

Table 10.1. (Continued from previous page). A modified PHREEQC input file for Example 11 from Parkhurst and Appelo (1999). The modifications are shown as commented out (lines beginning with #). The results are shown in Figure 10.1.

Example 11 of Parkhurst and Appelo (1999). The Example 11 input file is available from the PHREEQC distribution; this file plus all of the changes is listed in Table 10.1.

This brief description of the simulation is from Parkhurst and Appelo (1999), p. 238:

> The following example of advective transport in the presence of a cation exchanger is derived from a sample calculation for the program PHREEQM (Appelo and Postma, 1993, example 10.13, pp. 431–434). The chemical composition of the effluent from a column containing a cation exchanger is simulated. Initially, the column contains a sodium-potassium-nitrate solution in equilibrium with the cation exchanger. The column is then flushed with three pore volumes of calcium chloride solution. Calcium, potassium, and sodium react to equilibrate with the exchanger at all times. The problem is run two ways–by using the ADVECTION data block, which

models only advection, and by using the TRANSPORT data block, which simulates advection and dispersive mixing.

In PHREEQC, reactive transport is simulated by imagining a number of cells in a numbered sequence, containing minerals, surfaces, ion-exchangers, and an initial fluid equilibrated with these objects. The cells may all contain the same or different minerals and fluids. Another fluid is added to the first cell, equilibrated, with dispersion into the adjacent cell if desired, then that fluid is "shifted" or advected to the second cell, and the process is repeated until any number of "shifts" is performed.

In Figure 10.1, the ADVECTION option is illustrated in the left panel, and the TRANSPORT option in the right panel. The results from Parkhurst and Appelo's original Example 11 are shown in Figures 10.1(b) and (f). In Figures 10.1(c) and (g), the results of adding amorphous $Fe(OH)_3$ surfaces as modeled by Dzombak and Morel (1990) are shown.

Figures 10.1 (a), (b), (c), (e), (f), and (g) show the composition of the solution exiting from the column at cell 40, as a function of the amount of fluid that has flowed through, a total of three pore volumes. A pore volume is the cross-sectional area of the column times the length times the porosity, and the number of pore volume is equivalent to a dimensionless time. These diagrams, looking at the composition of the fluid exiting from the column versus time, are called *breakthrough curves*.

In Figures 10.1(d) and (h), we show the distribution of solutes along the whole column (cells 1–40) at a specific time, here after 40 shifts. These diagrams are called concentration *profile* diagrams.

Figures 10.1(a) and (e) represent the results for "non-reactive" advection (plug flow) and transport, i.e., advection plus dispersive mixing. There are no reactions between the solutes and aquifer solid matrix; all solutes are conservative in the simulations. Advection is the process by which the moving water carries solute with it at the same velocity. In the ADVECTION option, the $CaCl_2$ solution (called SOLUTION 0) is added to cell 1. This simply displaces the $(K, Na)NO_3$ solution to the next cell, and so on, as more SOLUTION 0 is added, until the $CaCl_2$ solution appears ("breaks through") at cell 40 after 1.0 pore volume. This indicates that the concentration fronts travel at the same velocity as the average linear velocity of the pore fluid. All the initial pore fluids in the column that contain Na, K, and NO_3^- are displaced by the infilling fluid. The column contains no more Na, K, or NO_3^-, and is entirely filled with the $CaCl_2$ solution. An abrupt interface between the infilling fluid and initial pore fluids is developed.

In Figure 10.1(e), the advective–dispersive transport results are shown. Mechanical dispersion occurs in the aquifer because groundwater is moving at rates that are both greater and less than the average pore velocity due to heterogeneities at various scales (Bear, 1972). As a result, mixing of fluids that have different solute concentrations occur, which tend to dilute the solute concentrations. In the TRANSPORT option, dispersive mixing takes place in addition to advection. As a result, a zone of mixing is created between the displacing fluid and the fluid being displaced (Figure 10.1(e)). Some mass of the solutes leaves the column in advance of the advective front (compared with Figure 10.1(a)) and some leave behind the advective front. The 0.5 C/C_o points for all the solutes coincide at the 1.0 pore volume for this one-dimensional transport problem.

In Figure 10.1(d), we show the distribution of solutes along the whole column (cells 1–40) after 40 shifts for pure advection (plug flow), showing that this is indeed the case. With added dispersive mixing (Figure 10.1(h)), a small amount of $(Na, K)NO_3$ solution persists near the exit cell, as also shown in Figure 10.1(e).

When heterogeneous reactions, i.e., reactions between solutes in groundwater and aquifer matrix, also occur in addition to advection and dispersion, the breakthrough curves for both `ADVECTION` and `TRANSPORT` options are quite different. The initial pore fluids are at equilibrium with the ion-exchanger X. In fact, PHREEQC calculates the equilibrium compositions of the ion-exchanger X from the given Na–K–NO$_3$ pore fluid. When the infilling fluid, with a chemistry as `SOLUTION 0`, invades the column, it brings an influx of $CaCl_2$ and moves out some $(K, Na)NO_3$. The equilibrium between groundwater and the exchanger is disturbed. To reach a new equilibrium, the following ion-exchange reactions proceed from left to right:

$$Ca + 2NaX \rightarrow CaX_2 + 2Na \qquad (10.15)$$

$$Ca + 2KX \rightarrow CaX_2 + 2K \qquad (10.16)$$

In the PHREEQC simulations, within each time step, the advective transport is calculated first, displacing the initial pore fluid with the invading $CaCl_2$ solution (called `SOLUTION 0`). This is followed immediately by a calculation of the chemical reactions with the exchanger. Here is where the static aqueous speciation and solubility modeling (Chapter 6) take place. The mass is re-distributed between the aqueous and solid phases according to equilibrium partitioning. The modeling is performed in the geochemical module. Then, dispersive transport is calculated using the mixing cell method. The disequilibrium brought about by the mass flux transported between spatial domains is again solved by the static speciation and solubility calculations. The mass transfer between the aqueous phase and solid phases are calculated from equilibrium partitioning. In this manner, the physical transport processes and chemical reactions are "coupled" through discretizing time and space into small increments so that the processes can be approximated by considering them separately.

We can see in Figures 10.1(b) and (f) that the concentration fronts are now "retarded" because of the reactions, compared with non-reactive transport. Na^+ in the column is completely displaced in about 1.5 pore volumes in the advective–reactive transport simulation versus 1.0 pore volume in the advective transport simulation. Because K^+ is bound to the exchanger more strongly, it is flushed out of the column later, at about 2 pore volumes. At this point, the column is saturated with Ca and the Ca front starts to appear in the effluent. Now due to ion-exchange reactions, the Ca front has traveled at about one-half of the average linear velocity of pore fluids. We can say Ca has a retardation factor of about 2 under these transport conditions.

The differences between Figures 10.1(b) and (f) are due to dispersion. Compared with advection–reaction transport, the sharp fronts in Figure 10.1(b) are "smeared" and the peak concentration of K is reduced. Note Cl^- is simulated as a conservative solute in all simulations.

When surface adsorption onto HFO is added in the reactive transport model, there is now competitive sorption of Ca^{2+} onto the ion-exchanger and HFO surfaces. Surface adsorption of the other four components in the model, Cl^-, NO_3^-, Na^+, and K^+, are not included in the Dzombak and Morel (1990) database. Therefore, we see the breakthrough curves for Ca^{2+}, Na^+, and K^+ depletion points arrive slightly later because a part of the Ca^{2+} mass influx is adsorbed on the HFO surfaces. The adsorption of Ca^{2+} displaces H^+ from the surfaces into solution and causes a decrease of solution pH.

A study of these results is worthwhile. Generally speaking, the addition of dispersive mixing "softens" or spreads out the results of pure advective flow, but sometimes the details can be surprising.

10.5   Buffering in Reactive Transport

10.5.1   The Buffer Concept

A buffered system is one that "resists" changes to its intensive variables. In other words, changing the composition of a buffered system will result in less change to certain properties than would occur in an unbuffered system. In chemistry, pH buffers are common. In geology, the buffer concept has been extended to many other intensive variables such as O_2 fugacity (Eugster, 1957), and HCl^o and HF^o activities (Zhu and Sverjensky, 1991).

Homogeneous buffers

The most familiar buffers, which we first learned about in high school chemistry, are mixtures of a weak acid and its salts. These are compositions that, when present in a solution, will use up or supply H^+, so that the pH is held close to some desired value when composition is changed, as during a titration.

For a solution of acetic acid and sodium acetate, for example, the following equilibrium holds:

$$HAc = Ac^- + H^+$$

where the Ac^- concentration is relatively high because of the NaAc. When acid is added to this solution, much of the added H^+ will be consumed by reaction with the acetate ion to form undissociated acetate acid (HAc). When a strong base is added to the solution, the added OH^- mostly reacts with H^+ to form water. More acetate acid then dissociates to replace the H^+ ions consumed by the reaction. As a result of the above reactions, the pH of the solution will be nearly constant as long as the acetate and acetate acid concentrations are not greatly altered (Rice and Ferry, 1982).

The $CO_2 - HCO_3^- - CO_3^{2-}$ equilibrium is just such a homogeneous buffer in natural waters. However, the buffer capacity is much smaller than that provided by heterogeneous buffers (Stumm and Morgan, 1996).

Heterogeneous Buffers

The most common heterogeneous buffer in the acidic and near neutral pH range in natural environments is calcite, the dissolution and precipitation of which can be represented by

$$CaCO_3 + H^+ = Ca^{2+} + HCO_3^-$$

When acid is added to a aquifer system that contains calcite, calcite will dissolve. The dissolution of calcite consumes H^+, which moderates the decrease of pH. The actual buffered pH value is determined by the Ca^{2+} concentration as well, which is, in turn, determined by many other chemical reactions. Thus, groundwater pH buffered by calcite can be different (though probably similar) at different sites.

Heterogeneous equilibria are the most efficient buffer systems for natural waters (Stumm and Morgan, 1996). They also represent a potential source of large buffer capacity, because in porous media the solid to water ratio is high.

Fixed Activity Buffers

In petrology, a buffered system usually refers to a system which is divariant, that is, which has the same number of components as phases, and which therefore has all its properties fixed at a given temperature and pressure. For example, a rock consisting of (pure) gypsum and anhydrite will have a fixed and invariable water activity at a given T and P. The relationship is

$$\tfrac{1}{2}CaSO_4 \cdot 2H_2O = \tfrac{1}{2}CaSO_4 + H_2O$$

and because the activities of the solid phases are 1.0, the equilibrium constant $K = a_{H_2O}$, meaning that the water activity cannot vary at that T and P, as long as both minerals are present. Adding water to the system will simply change the proportions of the solid phases. There are many such examples (cf. Rice and Ferry, 1982).

This is different in principle from the two pH buffers mentioned above. In these cases, the pH values do change upon addition of acid or base to the system, although only slightly, while, in a system with a fixed activity buffer, no properties can change at all at a given T and P, as long as the minerals are present, and as long as equilibrium prevails.

10.5.2 Application of the Buffer Concept

Reactive transport modeling is a realistic way of illustrating the origin of pH buffered zones at the Bear Creek site mentioned in Chapter 8 (§8.4). A conceptual model of the situation is shown in Figure 10.2. Sulfuric acid is pictured as flowing through a one-dimensional column filled with specific amounts of calcite, amorphous $Fe(OH)_3$ (or HFO), kaolinite, and quartz. Because of the varying solubilities of these minerals, they are successively dissolved away by the flowing acid, resulting in a pH zonation.

Titration of acid into a beaker containing these minerals will also successively dissolve away the minerals, as shown in §8.4. However, reactive transport modeling reveals a great deal more detail about the process, and has the capability of coming closer to reproducing natural processes, or, at the very least, of providing more insight into them. Such is the case here.

In the computer program input (Table 10.2), the column in Figure 10.2 is discretized into 40 cells, each 20 m long, and each containing specific amounts of calcite, kaolinite,

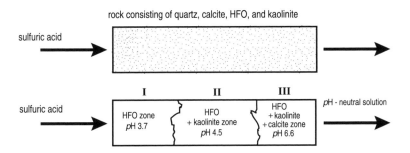

Figure 10.2. A conceptual experiment. A sulfuric acid solution passes through a rock consisting of HFO, kaolinite, calcite, and quartz. After a while, three pH zones are produced.

amorphous $Fe(OH)_3$ (HFO), and quartz. Gypsum is not present but is allowed to precipitate if it becomes supersaturated. A 0.1 molal solution of NaCl is equilibrated with this mineral assemblage in each cell. Conditions are oxidizing (equilibration with atmospheric O_2 and CO_2). Then into cell 1 is added 0.01 molal sulfuric acid (pH about 2), the resulting solution moved to cell 2, and so on until 200 shifts of solution have taken place. This means that 5 pore volumes of acid have flushed through the column.

The results of this simulation are numerous and varied; only a few are illustrated in Figure 10.3. Figures 10.3 (a), (c), (d), and (f) show results after 10, 50, 110, and 200 shifts (representing 13, 67, 147, and 267 years; or 0.25, 1.25, 2.75 and 5 pore volumes). Figure 10.3(b) shows only the result after 200 shifts. Figure 10.3(e) shows the results for several parameters after 200 shifts. In Figures 10.3(a), (d), (e), and (f), the Y-axis "moles" refers to the number of moles of mineral per liter present in the system. In Figure 10.3(b), the *change* in the number of moles of quartz per liter of groundwater between the shifts is plotted, as well as Saturation Index.

From the present point of view, the main conclusions are as follows.

- Not only do minerals dissolve behind the moving front, but they also precipitate ahead of the front, because the solution changes pH ahead of the flow front, and they become supersaturated. This cannot be simulated by the titration model.

- In the conceptual model (Figure 8.5) we imagined that the most acid zone would have only $Fe(OH)_3$ plus quartz. The results of the calculation, however, show that $Fe(OH)_3$ is completely removed from the zone closest to the inlet (Figures 10.3(a) and (e)). Thus, modeling can change our concepts, and direct us to things that should be further tested. Other input conditions could very likely be found for which a zone of only HFO plus quartz would be present.

- Comparison of Figures 10.3(b) and (d) shows that when kaolinite begins precipitating at about 400 m (at 267 years), quartz begins to dissolve to provide the necessary silica. Conversely, when kaolinite dissolves after 200 m, quartz becomes supersaturated and precipitates. These changes in the amount of quartz take place with no change in its Saturation Index, which has been given a value

```
TITLE Do buffer problem using reactive transport
      instead of titration
SOLUTION       0
        temp       25
        units      mol/kgw
        pH         2          charge
        S(6)       .01

SOLUTION       1-40
        temp       25
        units      mol/kgw
        ph         7
        Na         .1
        Cl         .1         charge
        S(6)       1e-7
        Si         1e-7
EQUILIBRIUM_PHASES     1-40
        CO2(g)     -3.5
        O2(g)      -0.7
        Fe(OH)3(a) 0.0        0.01
        Calcite    0.0        0.08
        Kaolinite  0.0        0.03
        Quartz     0.6        0.10
        Gypsum     0.0        0.0
END
TRANSPORT
# 20 m/shift, 15 m/y is 20/15 y/shift
# 31556925 s/y *20/15 y/shift = 42075900 sec/shift
# 200 shifts in 40 cells = 5 pore volumes
# 200 shifts * 1.333 y/shift = 267 y total flow time
        -cells         40
        -lengths       40*20
        -time          42075900
        -shifts        200
        -dispersivity  40*10
        -correct_disp  true
        -punch_cells   1-40
        -punch_frequency      10
PRINT
        reset              false
SELECTED_OUTPUT
        -file          buffering
        -simulation    false
        -solution      false
        -state         false
        -pe            false
        -si            Calcite Fe(OH)3(a) Kaolinite Gypsum Quartz
        -equilibrium_phases Calcite Fe(OH)3(a) Kaolinite Gypsum Quartz
END
```

Table 10.2. PHREEQC input for a reactive transport model to simulate fluid pH buffering at the Bear Creek site. This is a better alternative to the titration model in described Chapter 8.

of 0.6 (see input file, Table 10.2) to simulate the silica content of normal ground-waters, which are normally supersaturated with respect to quartz.

- Figure 10.3(e) shows the amounts present of each of three minerals, and how the pH is controlled by the mineral phases present. The middle pH zone is controlled by the presence of kaolinite. Also note that the pH is indifferent to the *amount* of kaolinite; it is the *presence* of kaolinite which is important.

- With a judicious choice of parameters, it should be possible to estimate the length of time it takes to develop the pH-buffered zones, and their position as a function of flow time. In this greatly simplified model, the pH values do not correspond well to the observed values, but the three zones are clearly developed.

10.6   Migration of an Acid Plume at a Uranium Mill Tailings Site

In Chapters 6 and 8, we introduced the Bear Creek Uranium site. Regulations require predictions and reports of concentrations over the next 200 years at the point of compliance. In 1994, a reclamation plan was proposed to install a low-permeability cover on the tailings ponds to prevent further infiltration from precipitation. Results from hydrological modeling show that tailings pore water will cease to drain into the underlying N sand 5 years after the cover installation. After that time, the plume will be flushed by uncontaminated upgradient groundwater. The distance of the migration of the acid plume and regulated metals and radionuclides will depend on the "natural attenuation" or the reactions of aquifer minerals with contaminated groundwater. The petitioner has to show to the Nuclear Regulatory Commission that the reclamation plan is adequate by showing that the concentrations of regulated chemicals do not exceed the limits. In reality, such predictions are almost always carried out by using K_d-based models. However, we show here that the coupled reactive transport model is more suited to simulate this chemically reactive and heterogenous system.

10.6.1   Model Description

The coupled reactive transport model is designed to simulate the acid plume migration under this "cover and attenuate" reclamation plan. An 800 m strip along cross-section A-A' (Figure 6.2) was discretized into 200 cells (Figure 10.4). Each cell is 4 m in length. The time step is 0.08 year.

A uniform and constant Darcy velocity of 15 m year^{-1} and effective porosity of 0.3 were used along the entire cross-section. The tailings fluid has relatively high chloride concentrations (0.016 mol L^{-1}) compared with the background water has only 0.0007 mol L^{-1}. It is widely believed that Cl$^-$ acts as a conservative solute in most aquifer systems, and thus its distribution can be used to retrieve dispersivity for the aquifer. By trial and error, a longitudinal dispersivity between 10 to 15 m appears to fit the concentration differences best in monitor wells sampled in September, 1994. It

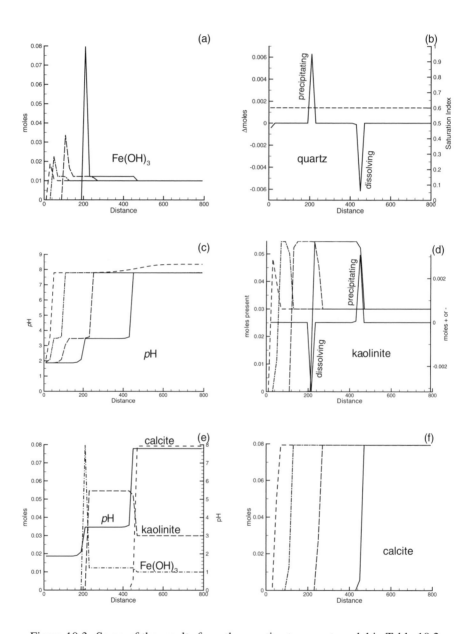

Figure 10.3. Some of the results from the reactive transport model in Table 10.2.

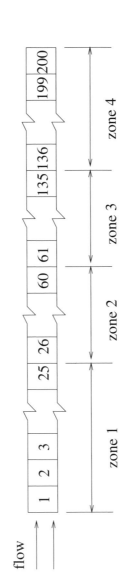

Figure 10.4. Discretization of the simulated domain. Numbers in the boxes are cell numbers.

is assumed in this study that molecular diffusion is negligible with respect to advection and dispersion.

From the field data and speciation–solubility modeling presented in Chapters 6 and 8, the distinct geochemistry zones in the aquifer could be interpreted with successive pH buffer reactions with calcite, amorphous $Al(OH)_3$, and amorphous $Fe(OH)_3$. This conceptual model was used as a geochemical submodel for the coupled reactive transport. A total of 11 aqueous components, H, Ca, Mg, Cl, CO_3^{2-}, Al, SO_4^{2-}, Fe^{3+}, Na, K, and Si, and six minerals, $Al(OH)_3(a)$, $Fe(OH)_3(a)$, calcite, gypsum, $SiO_2(a)$, and illite, were included in the simulations. For simplicity, chemical reactions were calculated at 25° C and 1 bar, although measured groundwater temperatures range from 12 to 16° C. The local equilibrium assumption (LEA) is used.

Initial conditions were specified to reflect site conditions in 1994 when a major field sampling program took place (Zhu *et al.*, 2002). The domain was divided into four zones (Figure 10.4), reflecting groundwater geochemical zonations observed in the field (Zhu *et al.*, 2002). The pore fluid chemistry and aquifer mineral compositions for each zone are shown in the PHREEQC input file (Table 10.3).

A flux (Cauchy or third type) boundary conditions are used for both ends of the one-dimensional strip. To represent the reclamation conditions, the incoming fluid has the chemistry of tailings pore fluid for the first five years and of uncontaminated upgradient groundwater thereafter (Table 10.3).

10.6.2 Modeling Results

As discussed in previous sections, coupled reactive transport models generate numerous data and the predicted mass transport is complex. Here, we only give some basic information from this modeling exercise. Interested readers are referred to Zhu and Burden (2001) and Zhu *et al.* (2001a, 2002) for details.

Distinct Zones and Sharp Fronts

Figure 10.5 shows the modeling results for the first five years. The first thing we notice is that the seepage of acidic tailings fluid into the shallow contaminated aquifer causes chemically distinct zones separated by sharp reaction fronts. Each zone corresponds to the successive buffer reactions with calcite, $Al(OH)_3(a)$, and $Fe(OH)_3(a)$.

$$CaCO_3 + 2H^+ = Ca^{2+} + H_2CO_3$$
$$Al^{3+} + 3H_2O = Al(OH)_3(a) + 3H^+$$
$$Fe^{3+} + 3H_2O = Fe(OH)_3(a) + 3H^+$$

The pH of groundwater is buffered to 6.3, 4.3, and 3.8, respectively.

Interactions Among Solutes

Results from this multi-component, multi-species transport model show that the transport of some components are in inter-locked step fashion. For example, the transport of sulfate is closely tied to calcite dissolution and precipitation.

```
TITLE    Transport: Bear Creek example in Phreeqc 2.0
SOLUTION 0   incoming solution
        units    ppm
        pH       3.8
        pe       15
        temp     25.0
        Ca       310.00
        Mg       1000.
        Cl       550
        C(4)     6.3
        Al       1020.
        S(6)     16500.
        Fe(3)    1950.0
        Na       360.00    charge
        K        60.
        Si       40.5
END
SOLUTION 1-25 Initial solution for column
units    ppm
        pH       3.8
        pe       15
        temp     25.0
        Ca       310.00
        Mg       1000.
        Cl       550
        C(4)     6.3
        Al       1020.
        S(6)     16500.
        Fe(3)    1950.0
        Na       360.00    charge
        K        60.
        Si       40.5
EQUILIBRIUM_PHASES 1-25
        Al(OH)3(a)   -1.36  0.0
        Fe(OH)3(a)    1.69  0.2
        Calcite       0.0   0.0
        Gypsum        0.0   0.2
        SiO2(a)       0.0   0.0
        Illite        0.0   0.0
SAVE EQUILIBRIUM_PHASES 1-25
SAVE Solution 1-25
END
```

Table 10.3. PHREEQC input for a reactive transport model. (Continued on next page.)

```
SOLUTION 26-60 Initial solution for column
        units    ppm
        pH       4.5
        pe       15
        temp     25.0
        Ca       420.00
        Mg       700.
        Cl       400
        C(4)     6.3
        Al       230.0
        S(6)     8100.00
        Fe(3)    926.00
        Na       278.00 Charge
        K        42.
        Si       10.
EQUILIBRIUM_PHASES 26-60
        Al(OH)3(a)  -1.36  0.1
        Fe(OH)3(a)   1.69  0.05
        Calcite      0.0   0.0
        Gypsum       0.0   0.2
        SiO2(a)      0.0   0.0
        Illite       0.0   0.0
SAVE EQUILIBRIUM_PHASES 26-60
SAVE Solution 26-60
END
SOLUTION 61-135 Initial solution for column
        units    ppm
        pH       6.5
        pe       15
        temp     25.0
        Ca       650.00
        Mg       250.
        Cl       375
        Alkalinity   1450
        Al       1.33
        S(6)     1650.00
        Fe(3)    2.18
        Na       212
        K        18.0
        Si       9.7
EQUILIBRIUM_PHASES 61-135
        Al(OH)3(a)  -1.36  0.01
        Fe(OH)3(a)   1.69  0.05
        Calcite      0.0   0.2
        Gypsum       0.0   0.0
        SiO2(a)      0.0   0.0
        Illite       0.0   0.0
SAVE EQUILIBRIUM_PHASES 61-135
SAVE Solution 61-135
END
```

Table 10.3. (Continued from previous page.) PHREEQC input for a reactive transport model. (Continued on next page.)

```
SOLUTION 136-200 Initial solution for column
        units   ppm
        pH      6.7
        pe      15
        temp    25.0
        Ca      550.00
        Mg      150.0
        Cl      275
        Alkalinity      878.
        Al      1.150
        S(6)    1500 charge
        Fe(3)   0.69
        Na      265.0
        K       14.
        Si      8.40
EQUILIBRIUM PHASES 136-200
        Al(OH)3(a)   -1.36    0.01
        Fe(OH)3(a)    1.69    0.05
        Calcite       0.0     0.2
        Gypsum        0.0     0.0
        SiO2(a)       0.0     0.0
        Illite        0.0     0.0
SAVE EQUILIBRIUM_PHASES 136-200
SAVE Solution 136-200
END
PRINT
        -reset  false
SELECTED_OUTPUT
        -file bearodu3.prn
        -totals H Cl Fe Al S Na Mg C Ca K Si
        -equilibrium_phases Al(OH)3(a) Fe(OH)3(a) Calcite Gypsum SiO2(a) Illite
        -pe false
        -time true
        -step false
        -state false
TRANSPORT
        -cells          200
        -length         4
        -shifts         62
        -time_step      2522880
        -flow_direction forward
        -boundary_conditions  flux  flux
        -correct_disp   true
        -dispersivity   10.0
        -diffusion_coef 0.0e-0 #No diffusion
        -tempr 1.0      # No heat retardation
END
SOLUTION 0  Flush with background groundwater
        units   ppm
        pH      7.4
        temp    25.0
        Ca      158.00
        Mg      21.00
        Mn(3)   0.11
        Cl      25.     charge
        C(4)    153
        Al      .0100
        S(6)    425.00
        Fe(3)   0.1
        Na      61.0
        K       7.
        Si      5.60
END
SELECTED_OUTPUT
        -file bearout.prn
        -totals Cl Fe Al S Na Mg C Ca K Si
        -equilibrium_phases Al(OH)3(a) Fe(OH)3(a) Calcite Gypsum SiO2(a) Illite
        -pe false
        -time true
        -step false
        -state false
TRANSPORT
        -shifts 2500
END
```

Table 10.3. (Continued from previous page.) PHREEQC input for a reactive transport model to simulate plume migration at the Bear Creek site.

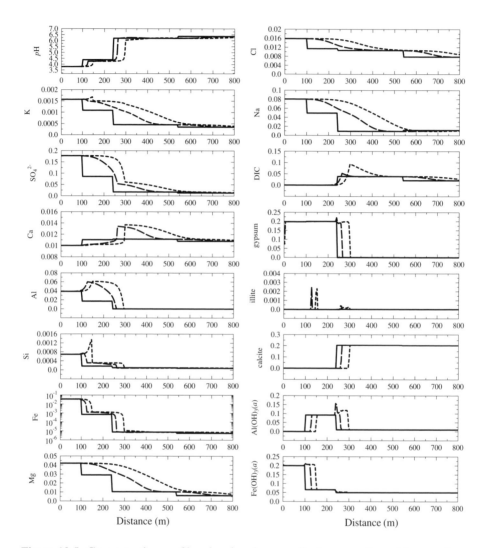

Figure 10.5. Concentration profiles showing downgradient propagation of the reaction fronts and pH-buffered geochemical zones during tailings fluid intrusion. Groundwater flows from left to right. DIC denotes total dissolved inorganic carbon. All concentration units are in mol L^{-1}, except for pH. Solid, dashed, and dotted lines represent concentration profiles at 0 (initial conditions), 2.5, and 5 years from the start of the simulation, respectively.

The rise of pH from about 4.3 to 6.3 because of calcite dissolution also causes the precipitation of gypsum (Figure 10.5),

$$CaCO_3 + SO_4^{2-} + H^+ + 2H_2O = CaSO_4 \cdot 2H_2O + HCO_3^- \qquad (10.17)$$

During the flushing period, the uncontaminated upgradient groundwater dissolves gypsum. Dissolution of gypsum causes precipitation of a small amount of calcite and that in turn causes a drop of pH. The drop of pH causes dissolution of small amounts of $Al(OH)_3(a)$ and $Fe(OH)_3(a)$.

Multiple Concentration Waves

The coupled reactive transport model predicts multiple concentration waves. Helfferich (1989) defined "wave" as any variation of solute or solid phase concentration and is synonymous with "front". Multiple concentration waves are evident in the sulfate breakthrough curves (BTC) at the 200th cell near the property boundary. In Figure 10.6, we can see two SO_4^{2-} concentration waves, one associated with the advection–dispersion arriving at about 1 pore volume and another with gypsum dissolution at about 11 pore volumes. The first wave or concentration peak is equal to the advection speed. After about 11 pore volumes, all gypsum is expended. The sulfate concentration drops sharply and then returns to background levels in about 172 years. Between the two fronts, a concentration plateau is maintained by gypsum equilibrium at the upgradient, conforming to the "downstream equilibrium condition" (Walsh *et al.*, 1984).

A Case Against K_d-based Transport Models

In most practical applications, concentration predictions are from K_d-based "reactive" transport models. For comparison purposes, we used the one-dimensional finite difference code Bio1D (Srinivasan and Mercer, 1987) for transport simulation. The code can simulate advective–dispersive–reactive transport and biodegradation. In our simulation, all transport parameters and initial and boundary conditions were set to resemble the coupled transport model as closely as possible. The results using different values of K_d are shown in Figure 10.6(b).

From the coupled reactive transport modeling results, the shortcomings of the K_d approach are obvious:

- K_d models cannot predict the sharp fronts and distinct geochemical zones (cf. Figure 10.5);

- K_d models can only simulate the transport of one solute at a time, whereas the coupled model shows inter-locking interactions and transport among the solutes (cf. Figure 10.5);

- K_d models can only simulate one concentration wave, whereas multiple concentration waves are obvious in this case (cf. Figures 10.5 and 10.6a).

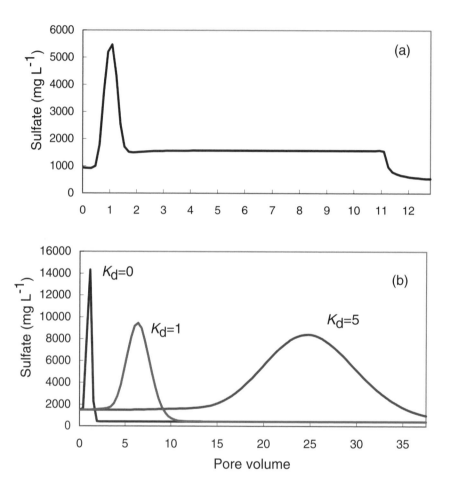

Figure 10.6. Breakthrough curves for sulfate at the 200th cell from (a) the coupled reactive transport model (surface adsorption reactions are added) and (b) the K_d-based model.

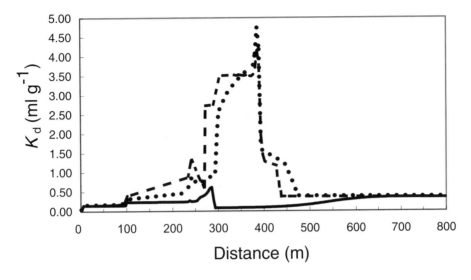

Figure 10.7. K_d values calculated from results of coupled reactive transport simulation. Solid line, 0 year flushing; dashed line, 100 year; dotted line, 200 year.

Further, we can show that K_d values vary both spatially and temporally. For example, effective K_d values for sulfate can be calculated from simulation results based on the following equation:

$$K_d = \frac{(C_{\text{gypsum}} + C_{\text{surface}})}{C_{t,aq}} \frac{\theta}{\rho_b} \qquad (10.18)$$

where C denotes concentrations in mol L^{-1}, and $C_{t,aq}$ denotes total dissolved sulfate concentrations. In these calculations, we used the results from simulations that included the surface adsorption reactions.

Figure 10.7 shows the K_d (ml g^{-1}) distribution at the end of 5 years of tailings water intrusion and at 100 and 200 years of flushing with uncontaminated groundwater. K_d varies spatially from 0.08 to 4.88 (ml g^{-1}), a 60-fold change. More significantly, K_d values change with time. The changes of calculated K_d are related to the chemical reactions that determine the partitioning of SO_4^{2-} between the solid matrix and groundwater as discussed above.

10.7   Remedial Design of a Uranium Tailings Repository

Morrison *et al.* (1995b) simulated reactive transport of uranium through a liner/reactive barrier containing adsorptive hydrous ferric oxide (HFO). Their simulations were intended to evaluate the effectiveness of a preliminary engineering design for a repository that would contain $2.3 \times 10^6 \text{ m}^3$ of uranium mill tailings and contaminated soil at Monticello, Utah. The proposed repository will be excavated in pediment gravels that are underlain by the relatively impermeable Mancos Shale. The only remediation measure will be an engineered layer of re-packed pediment gravel containing HFO, which will

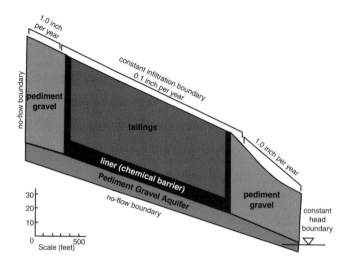

Figure 10.8. Repository design and boundary conditions for the flow model.

act as a chemical barrier. The HFO will be produced by spraying $FeCl_3 \cdot 6H_2O$ and assuming it will react with calcite in the pediment gravel to form HFO. The tailings to be placed into the new repository will contain water because of original pore water, infiltration from precipitation, and water added to control dust and compaction. It is estimated that initially the tailings will be >80% saturated. A low-permeability earthen cover, about 3 m thick, will be placed over the tailings repository and will be vegetated. Therefore, release of contaminants will be due to the draining of pore fluids over time, mainly in a lateral direction because of the impermeable nature of the underlying Mancos Shale. The regulatory standard at the site for uranium concentrations is 30 pCi per liter or $0.05 \, \text{mg} \, \text{L}^{-1}$ in the effluents. The compliance period is 200 years.

Morrison *et al.* (1995b) used the coupled reactive transport model HYDROGEOCHEM developed by Yeh and Tripathi (1991) to simulate uranium transport in a two-dimensional vertical cross-section of the site, approximately along a streamline (Figure 10.8). The flow field needed for the reactive transport modeling was simulated using HYDROFLOW, a simplified version of program FEMWATER. Figure 10.8 shows the boundary conditions for the unsaturated–saturated flow model.

Chemical equilibrium conditions are assumed for all reactions. The chemical model consists of eight components UO_2^{2+}, VO_4^{3-}, CO_3^{2-}, K^+, Ca^{2+}, H^+, and HFO as the sorbent. Four minerals, carnotite ($K_2(UO_2)_2(VO_4)_2$), tyuyamunite ($Ca(UO_2)_2(VO_4)_2$), calcite ($CaCO_3$), and gypsum ($CaSO_4 \cdot 2H_2O$), are allowed to participate in precipitation-dissolution reactions. The detailed chemical model (reactions and parameters) is presented in Morrison *et al.* (1995a). No ionic strength correction was made for activity coefficients.

The initial conditions for the transport model were the total analytical concentrations of components estimated from the complete digestions of tailings samples. Four tailings piles exist at the site and will be filled in the repository: Carbonate, Vanadium, Acid and East Piles. The Vanadium Pile contains pore fluids with *p*H values from 7 to 13 and

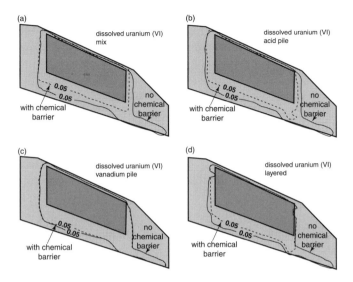

Figure 10.9. Position of the $0.05 \ \mathrm{mg \, L^{-1}}$ dissolved uranium contour after 11 years. (a) All "Mix" tailings; (b) Acid Pile tailings; (c) Vanadium Pile tailings; and (d) 20% Vanadium Pile on top and 80% Acid Pile at the bottom. Shading shows the location of tailings.

alkalinity up to $9\,000 \ \mathrm{mg \, L^{-1}}$ as $CaCO_3$. The "Acid Pile" pore fluids range from 7 to 8.5. Pore fluids in the East Pile and Carbonate Pile have pH values from 7 to 8. During simulations, one scenario calculated the placement of "Mix", a composite tailings with equal amounts from all the four tailings piles.

The total concentration of the sorption sites was estimated from the results of citrate–bicarbonate–dithionite extraction of iron oxides (Smith and Mitchell, 1987). For HYDROGEOCHEM, the total concentration of a component includes concentrations of a component in the solid phases, the aqueous phase, and adsorbed species on the surfaces. For a given input of the total concentrations, the equilibrium modeling module, EQMOD, distributes the total concentrations among the aqueous solution, solid phases, and mineral surface sites according to the reactions and thermodynamic parameters provided in the input file. Morrison *et al.* (1995b) found that the calculated pore fluid chemistry from EQMOD is different from the analyses of the pore fluid samples. To be more consistent with the pore fluid chemistry, they adjusted the total analytical concentrations in the model to within 10% of the total digestion concentrations, and assigned low concentrations to the pediment gravel and assumed that the gravel is not reactive. The sorption sites in the chemical barriers were calculated to be equivalent to $1\,000$ tons of raw material of $FeCl_3 \cdot 6H_2O$ total in the entire repository.

The boundary concentrations for the transport model are the same low-concentrations as in the pediment gravel. Oxidizing conditions were assumed to exist throughout the system (Eh from 400 to 500 mV). Only advection was considered, and dispersivity was set to zero. A period of 216 years was simulated.

Figures 10.9 and 10.10 show the simulation results. The simulations suggest that

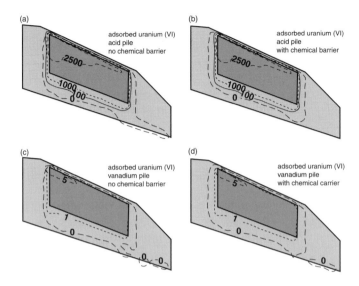

Figure 10.10. Distribution of adsorbed uranium after 11 years (in $mg\,L^{-1}$). Shading shows the location of tailings.

installing a chemical barrier can significantly retard the uranium migration for all the tailings piles except the Vanadium Pile. Assuming the tailings chemistry is similar to the mixed sample, this chemical barrier can effectively limit the migration of the 0.05 $mg\,L^{-1}$ contour to within 27 m of the repository boundary in 216 years.

The coupled reaction transport modeling effort provided a few insights for the design of remedial action. It was found that:

- For the mixed tailings, the 0.05 $mg\,L^{-1}$ front, the regulatory threshold, will migrate 27 m from the repository boundary in 216 years;

- A chemical barrier of HFO liner is effective in retarding migration of U(VI) from near-neutral pH uranium mill tailings, but not from alkaline mill tailings;

- The simulation suggests that piling near neutral pH tailings into the repository first and then adding the alkaline tailings on the top would help increase the adsorption in the chemical barrier and reduce uranium migration;

- If the chemical barrier is placed more in the zones of highest groundwater flow, not evenly distributed, the performance would be improved.

Morrison *et al.* (1995b) also discussed the model uncertainties. They cited the lack of accurate thermodynamic properties, the lack of information on tailings mineralogy, and the difficulty of characterizing and representation of site heterogeneities.

10.8   Summary and Comments

As we discussed earlier, the need for predictions of spatial and temporary concentration distributions is driven by regulatory requirements. For the example of the uranium mill tailings impoundment discussed above, in order to receive approval for the "cover and naturally attenuate" reclamation plan, the license applicant needs to show to the regulatory agency that natural attenuation can effectively reduce the contaminant concentration levels so that the licensed groundwater constituents will not exceed regulatory levels at the point of compliance.

The shortcomings of the empirical K_d- and isotherm-based models are well known and obvious. Nevertheless, nearly all computer models of fate and transport in the regulatory environment use this approach. Decisions for thousands of environmental cases and millions of dollars spent are based on predictions from empirical models.

The acceptance of the time-honored retardation-factor approach has enormously compromised the role of geochemistry and geochemists in the environmental field. A geochemist's work has been reduced to supplying a K_d value to the hydrogeologist. Understanding of site geochemistry has been reduced to simple column and batch experiments to obtain a K_d value. The need for careful characterization of mineralogy has never become an issue, while millions are spent on analyzing groundwater compositions because the geochemical heterogeneity factor is never entered into the transport equation. At least one advantage of coupled reactive transport models is to put geochemistry back on the front burner.

While the "mechanistic" treatment of chemical reactions in the coupled multi-component, multi-species mass transport has obvious advantages over the empirical isotherm-based transport models, we can also easily compile a long list of shortcomings for coupled reactive transport models. We choose a few and list them here.

1. Coupled models are computationally expensive. In order to develop a "manageable" model, the expansion of chemical reactions often comes at the expense of physical hydrogeology. Commonly, a uniform and constant Darcy velocity was assumed (e.g., Zhu *et al.*, 2001a). Often, the transport problem is reduced to one or two dimensions.

2. A coupled model can easily generate so much simulation data that the comprehension of chemical reactions is difficult. It would require many more man-hours to fulfil regulatory requirements.

3. Most applications of coupled models use the local equilibrium assumption. It is well known that most heterogeneous and redox reactions in the low temperature, near-surface systems are kinetically controlled (Hunter *et al.*, 1998). However, we lack both theory and data to model kinetic reactions satisfactorily. In addition, inclusion of kinetic reactions makes the results even more difficult to comprehend completely and hence more costly.

4. There is a general lack of mineralogical studies in site characterizations (Zhu and Burden, 2001). Actual mineral assemblages, both primary and secondary, their relative abundances, and spatial distributions are very important to the modeling results. However, detailed mineralogical data are seldom available.

11

Kinetics Modeling

11.1 Introduction

Because our general aim is to illustrate geochemical modeling as practised at the present time, our goal in this chapter is limited by the capabilities of programs (widely) available today. Therefore, we do not attempt to introduce the broad field of chemical kinetics, but only the amount necessary to understand the kinetic capabilities of today's modeling programs. We give examples using what we consider to be the most useful of these, The Geochemist's Workbench™ and PHREEQC. As mentioned earlier, our treatment is no substitute for a careful reading of the user's manuals for these programs.

11.2 Some Basic Theory

We need a mathematical expression involving the concentrations of reactants and products in a given chemical reaction, which relates how far the reaction has proceeded to some time variable. First, we introduce a variable which tells us how far the reaction has proceeded.

11.2.1 The Progress Variable

Consider a generalized chemical reaction

$$a\text{A} + b\text{B} = c\text{C} + d\text{D} \tag{11.1}$$

where A, B, C, and D are chemical formulae, and a, b, c, d are the stoichiometric coefficients. We pointed out in §3.4.3 that when this reaction reaches equilibrium,

$$c\mu_\text{C} + d\mu_\text{D} = a\mu_\text{A} + b\mu_\text{B}$$

and

$$\Delta_r\mu = c\mu_\text{C} + d\mu_\text{D} - a\mu_\text{A} - b\mu_\text{B} \tag{11.2}$$
$$= 0$$

230

Equation (11.1) can be generalized to

$$\sum_i v_i M_i = 0 \qquad (11.3)$$

where M_i are the chemical formulae and v_i are the stoichiometric coefficients, with the stipulation that v_i is positive for products and negative for reactants. Equation (11.2) can then be generalized to

$$\sum_i v_i \mu_i = 0 \qquad (11.4)$$

Up to now, we have been most interested in reactions that reach equilibrium. Now let's look at what happens before that point is reached, i.e., while the reaction is taking place. Let's say that the reaction proceeds from left to right as written. It doesn't matter for the moment whether all the reactants and products are in the same phase (a *homogeneous* reaction) or in different phases (a *heterogeneous* reaction). During the reaction, A and B disappear and C and D appear, but the *proportions* of A:B:C:D that appear and disappear are fixed by the stoichiometric coefficients. If the reaction is

$$A + 2B \rightarrow 3C + 4D \qquad (11.5)$$

then for every mole of A that reacts (disappears), 2 moles of B must also disappear, while 3 moles of C and 4 moles of D must appear. This is simply a mass balance, independent of thermodynamics or kinetics, and can be expressed as

$$\frac{dn_A}{v_A} = \frac{dn_B}{v_B} = \frac{dn_C}{v_C} = \frac{dn_D}{v_D} = \frac{dn_i}{v_i} \qquad (11.6)$$

where the differentials dn_A, dn_B, and so on, refer to a change in the amount of A, B, and so on, of any convenient magnitude. We can next imagine the reaction proceeding in a series of such changes, or reaction increments, from the beginning to the end of the reaction (either when equilibrium is reached, or when one of the reactants is used up), and write

$$\frac{dn_A}{v_A} = \frac{dn_B}{v_B} = \frac{dn_C}{v_C} = \frac{dn_D}{v_D} = d\xi \qquad (11.7)$$

from which it appears that

$$\frac{dn_A}{d\xi} = v_A; \quad \frac{dn_B}{d\xi} = v_B; \quad \cdots \frac{dn_i}{d\xi} = v_i \qquad (11.8)$$

where our new variable ξ is called the reaction progress variable, and in this case represents an arbitrary number of moles. Equation (11.8) says that in Reaction (11.5), $dn_A/d\xi = -1$, $dn_B/d\xi = -2$, $dn_C/d\xi = 3$, and so on, which simply means that for every mole of A that disappears, 2 moles of B also disappear, 3 moles of C appear, and so on.

11.2.2   The Reaction Rate

Having defined reaction increments $d\xi$, we can now define the *rate of reaction* as

$$\frac{d\xi}{dt} = \frac{1}{\nu_A}\frac{dn_A}{dt} = \frac{1}{\nu_B}\frac{dn_B}{dt} = \cdots = \frac{1}{\nu_i}\frac{dn_i}{dt} \qquad (11.9)$$

where dt is an increment of time, and $d\xi/dt$ is the derivative of ξ with respect to t, an expression of the amount of progress of the reaction as a function of time, or simply the rate of reaction.

This expression (11.9) is written in terms of the absolute number of moles of A, B, and so on, $(n_A, n_B, \dots)$, but by considering a fixed volume we could change these to concentration terms. Thus,

$$\frac{d\xi}{dt} = \frac{1}{\nu_A}\frac{dC_A}{dt} = \frac{1}{\nu_B}\frac{dC_B}{dt} = \cdots = \frac{1}{\nu_i}\frac{dC_i}{dt} \qquad (11.10)$$

where C is in some unit of concentration such as moles per cubic centimeter.

So, evidently, the rate of reaction can be determined by measuring the concentration of *any* of the reactants or products as a function of time. With one important stipulation: the reaction we have written must be what is actually happening.

Elementary and Overall Reactions

If we measure the rate of change of concentration of products and reactants in many ordinary chemical reactions, we find that the relationship in (11.10) is often not obeyed. This is because the reaction does not actually proceed *as written*, at the molecular level. For example, Reaction (11.5), taken literally, indicates that a molecule of A reacts with 2 molecules of B, and at that instant, 3 molecules of C and 4 molecules of D are formed. But this might not be what happens at all, and in view of the improbability of three molecules (A + 2 B) meeting at a single point, it probably is not in this case. The reaction as written may well represent the *overall* result of a series of *elementary* reactions. Thus A and B may in fact react to form a number of *intermediate species* such as X and Y, which then react with each other or with A or B to form C and D. In thermodynamics, the existence of such intermediate species is not important to the study of the overall reaction, as long as equilibrium is attained, but, in kinetics, these intermediate species contribute to the overall rate of reaction and may actually be *rate-controlling*, even though their concentrations may be small.

The *reaction mechanism* is the description of an overall reaction in terms of the separate elementary reactions that are involved.

The Steady State

Of course, it is also possible that intermediate species do form, but they achieve a *steady-state* concentration, that is, they break up just as rapidly as they form. In this case, Equation (11.10) would be obeyed, even though it did not represent what actually happens at the molecular level. Steady-state conditions are common in experimental and natural settings. For example, the CO_2 concentration in the atmosphere is not controlled by any equilibrium reaction, but is the net result of a balance between inputs and outputs from a variety of processes. The "steady-state" concentration may of course gradually change, but so slowly compared with other rates in the system that considering it a constant in kinetic models is correct (Lasaga, 1998, §1.3).

11.2.3   Rate Laws

A rate law is a statement about how the rate of a reaction depends on the concentrations of the participating species. If one thinks about chemical reactions as something that happens at the molecular level when molecules collide with one another, it makes sense that the number of collisions, and hence the rate of reaction, should depend on how many molecules of each type there are; that is, their concentrations.[1]

In most cases, a simple power function of concentrations is found to apply. For reaction (11.1) it is

$$\text{rate of reaction} = R = \frac{d\xi}{dt} = k \cdot C_A^{n_A} C_B^{n_B} C_C^{n_C} C_D^{n_D} \tag{11.11}$$

The constant of proportionality, k, is called the *rate constant*. The exponents $n_A \ldots n_D$ are often integers, but can be fractional or decimal numbers, especially in heterogeneous reactions where adsorption and other surface-related effects can influence reaction rates. They define the *order* of the reaction. If n_A is 2, the reaction is said to be second order in A. The sum of the exponents gives the overall order of the reaction.

If the reaction is heterogeneous (more than one phase is involved), one or more of the "concentrations" in (11.11) must refer to the *specific area* (surface area per unit volume of solution) or the *reactive* surface area of the solids involved, because obviously the rate of reaction will depend on how much of the solid is available to react. This is one of the most difficult parameters to quantify, especially in field situations.

Simple Rate Equations

Rate equations or rate laws are determined by analyzing one or more reactant or product species as a function of time as a reaction proceeds, and then inspecting the results to see what theoretical form best fits the data. The simplest examples are those for elementary reactions, but the kinetics of more complex (multi-step) reactions can often be expressed by the same equations in certain cases, such as by using suitable constraints.

Zeroth Order   A zeroth-order reaction obeys the relation

$$-\frac{dC}{dt} = k \tag{11.12}$$

or

$$C = C^\circ - kt \tag{11.13}$$

---

[1] In thermodynamics, we must use "corrected" concentrations, or activities. In kinetics it is the actual concentrations that are important.

where $C°$ is the initial concentration. A plot of concentration vs. time will thus give a straight line. Bacterial reduction of sulfate in marine environments (where variables other than SO_4^{2-} are constant) appears to be zeroth order (Lasaga, 1998, p. 38).

First Order Rate laws for elementary reactions are for the most part what one would expect. For example, a simple molecular (or nuclear) decomposition,

$$A \rightarrow products$$

proceeds at a rate that depends only on the concentration of A; the more A, the more decomposition per unit time. If all exponents in Equation (11.11) are zero except for $n_A = 1$, the rate law becomes

$$\frac{d\xi}{dt} = -\frac{dC_A}{dt} = k \cdot C_A \tag{11.14}$$

and the reaction is first order. The decay of radioactive elements is an example of such reactions.

If we simplify C_A to C and let $C = C°$ at time $t = 0$, integration of (11.14) gives

$$\int_{C°}^{C} \frac{dC}{C} = -k \int_0^t dt \tag{11.15}$$

$$\ln \frac{C°}{C} = kt \tag{11.16}$$

$$\ln C = \ln C° - kt \tag{11.17}$$

$$C = C°e^{-kt} \tag{11.18}$$

These equations suggest various ways of plotting data to see if they fit a first-order rate law. For example, a plot of $\ln(C°/C)$ vs. t will give a straight line with a slope equal to the rate constant for concentration data from a first-order reaction.

Second Order The most common type of elementary reaction resulting from bimolecular collisions is

$$A + B \rightarrow products \tag{11.19}$$

Here we expect the frequency of reaction to be proportional to the concentrations of the reactants and the concentration of products to have no effect, and so the rate law is

$$\frac{d\xi}{dt} = -\frac{dC_A}{dt} = -\frac{dC_B}{dt} = k \cdot C_A^1 C_B^1 \tag{11.20}$$

and the reaction is second order.

If the initial concentrations of A and B are $C_A°$ and $C_B°$, the stoichiometry of (11.19) requires that

$$C_A° - C_A = C_B° - C_B$$

Solving this for C_B and substituting this result in (11.20) yields

$$-\frac{dC_A}{dt} = k \cdot C_A(C_A - C_A° + C_B°) \tag{11.21}$$

Integration of (11.21) then gives

$$\ln\left(\frac{C_A^{\circ} C_B}{C_B^{\circ} C_A}\right) = (C_B^{\circ} - C_A^{\circ})kt \qquad (11.22)$$

Therefore, a plot of

$$\left(\frac{1}{C_B^{\circ} - C_A^{\circ}}\right)\ln\left(\frac{C_A^{\circ} C_B}{C_B^{\circ} C_A}\right)$$

vs. time t will result in a straight line with a slope equal to the rate constant for concentrations taken from a second-order reaction. Similar equations can be derived for reactions with different stoichiometric coefficients.

There are a number of other rate laws, but this will suffice to give an idea of the procedures involved. However, it should be emphasized that most chemical reactions are "overall" reactions, and that understanding them in terms of their fundamental elementary reactions is a goal not often and not easily achieved.

Pseudo-Order Reactions   As mentioned above, complex reactions can often be expressed by the simple equations of zeroth-, first-, or second-order elementary reactions under certain conditions. For example, the dissolution of many minerals at conditions close to equilibrium is a strong function of the free energy of the reaction (Lasaga, 1998, §7.10), but far from equilibrium the rate becomes nearly independent of the free energy of reaction. In other words, the rate of dissolution will be virtually constant under these conditions, or pseudo-first-order.

Another example of a pseudo-first-order reaction is shown in Figure 11.1. These data are the result of a very complicated set of reactions, largely unknown, involving non-bacterial reduction of aqueous sulfate and slow release of hydrogen sulfide from organic material in a natural shale. Nevertheless, the data fit a simple first-order relationship.

In a reaction involving A and B which is truly second order, choosing conditions such that the concentration of B is held essentially constant (which might be done in several ways) reduces Equation (11.20) for second-order reactions to

$$-\frac{dC_A}{dt} = kC_B C_A = k_{app} C_A \qquad (11.23)$$

which is the equation for a first-order reaction, where $k_{app} = kC_B$, is a pseudo-first-order rate constant, or apparent-first-order constant. If *all* reactant concentrations are held essentially constant, a pseudo-zeroth-order reaction is generated.

11.2.4 Temperature Dependence of Rate Constants

From the relations in §3.5.3 we see that the temperature dependence of the equilibrium constant K can be expressed as

$$\frac{d \ln K}{d(1/T)} = -\frac{\Delta_r H^{\circ}}{R}$$

or, alternatively,

$$\frac{d \ln K}{dT} = \frac{\Delta_r H^{\circ}}{RT^2}$$

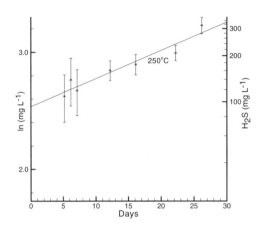

Figure 11.1. Concentration of H_2S in a solution in contact with gypsum and organic-rich shale at 250°C, as a function of time. The H_2S is being released from the organic shale, and is also being formed by reduction of aqueous sulfate (unpublished data of G.M. Anderson and T. M. Seward).

In 1889, Arrhenius proposed a similar equation for the temperature effect on rate constants,

$$\frac{d \ln k}{d(1/T)} = -\frac{E_a}{R} \qquad (11.24)$$

or

$$\frac{d \ln k}{dT} = \frac{E_a}{RT^2} \qquad (11.25)$$

where E_a is the Arrhenius activation energy, or just the activation energy, and turns out to be closely related to the "energy barrier" between products and reactants in chemical reactions.

Experimental data for a great many reactions over a large range of temperatures show that the Arrhenius equation is usually closely obeyed, showing that E_a is either a constant or a weak function of temperature, and so we can integrate the equation to give

$$\ln k = \ln A - \frac{E_a}{RT} \qquad (11.26)$$

or

$$k = Ae^{-E_a/RT} \qquad (11.27)$$

where A, which enters (11.26) as a constant of integration, is called the *pre-exponential factor*.

The activation energy of an overall reaction is made up of the individual contributions of the elementary reactions making up the overall reaction. The magnitude of the activation energy can vary from virtually zero to hundreds of kilojoules per mole and, besides controlling the temperature dependence of the rate constant, provides clues as to the nature of the reaction mechanisms, because the energies involved in many types of diffusion, electron exchange, and bond-breaking processes are known.

11.3   Kinetics of Precipitation and Dissolution Reactions

A reasonable assumption about the rate of mineral precipitation or dissolution might be that the rate is proportional to the degree of super- or under-saturation. Expressing this idea mathematically, we would say that the rate R is

$$R \propto (1 - \Omega) \tag{11.28}$$

where Ω is defined as the ratio of the ion activity product (IAP) and the solubility product, K_{sp} (§3.5.2). If IAP $< K_{sp}$, the solution is underaturated, $\Omega < 1$, and R is positive (mineral dissolves). If IAP $> K_{sp}$, $\Omega > 1$, and R is negative (mineral precipitates). For added generality, we would use Q and K rather than IAP and K_{sp} (§3.5; §3.5.2).

The proportionality factor would include, at a minimum, the rate constant for the reaction, and the relative surface area of the mineral, i.e., the mineral surface area per unit volume of solution. We might also add a term, ν, to account for cases in which the dissolution of one mole of solid phase gives rise to ν moles of the component we are measuring. For example, if we measured the solubility of Mg_2SiO_4 by measuring aqueous Mg, $\nu = 2$.

Equation (11.28) then becomes

$$R = \frac{A}{V}\nu k \left(1 - \frac{Q}{K}\right) \tag{11.29}$$

where A and V are the mineral surface area and solution volume, respectively, and k is the rate constant of the dissolution reaction. Other options are the inclusion of an exponent n on the $(1 - \Omega)$ term, i.e., $(1 - \Omega)^n$, or on the Ω term itself, i.e., $(1 - \Omega^n)$. This equation is used in The Geochemist's Workbench™ to model the kinetics of dissolution and precipitation. Program PHREEQC uses a more flexible method, as explained below.

Examples

The Geochemist's Workbench™   In program REACT, part of The Geochemist's Work-bench™, equation (11.29) is implemented by using the `kinetic` command to set the variables (key words) required. The format is

```
kinetic <mineral> <variable> = <value>
```

where `mineral` is an identifying label for the following variable, and `variable` is either `rate_con` for the rate constant in moles per square centimeter per second, or `surf` for the mineral surface area in square centimeters per gram.

For example, the rate constant for the dissolution of albite is $10^{-12.26}\,mol\,m^{-2}\,s^{-1}$, or $10^{-16.26}\,mol\,cm^{-2}\,s^{-1}$ (Lasaga, 1998, Table 1.5). This is entered in REACT as

```
kinetic Albite rate_con 5.50e-17
```

```
T = 25
swap Kaolinite for Al+++
swap Quartz for SiO2(aq)
swap Muscovite for K+
pH = 6
Na+   = .3  molal
Ca++  = .05 molal
Cl-   = .3  molal
HCO3- = .02 molal
 10 free grams Kaolinite
 10 free grams Quartz
 10 free grams Muscovite
react 10 grams Albite
kinetic Albite pre-exp 18370 act_en 117200
kinetic Albite surf = 1e3
time begin 0 years, end 2000 years
suppress all
unsuppress Albite Quartz Muscovite Kaolinite
go
```

Table 11.1. Input script for program REACT, to model the dissolution of albite.

Alternatively, we can set the activation energy (E_a) and the pre-exponential factor (A), Equation (11.27), to let REACT calculate the rate constant at any temperature. Thus

```
kinetic Albite pre-exp 18370 act_en 117200
```

(using values from Lasaga, 1998).

To use a surface area of $1000 \text{ cm}^2 \text{ g}^{-1}$ (the solution volume V is known by the program), we enter

```
kinetic Albite surf 1000
```

The program then has enough information to use Equation (11.29). However, if some other rate equation is preferred, we can enter it in another part of the kinetic statement (see below). For example, to use Equation (11.29) to examine the kinetic dissolution of albite in water in equilibrium with kaolinite, muscovite, and quartz at 25°C and a pH of 6.0,[2] we would prepare a script similar to that shown in Table 11.1.

We suppress the effects of all minerals except those needed, to allow a better comparison with PHREEQC (below), with the suppress all and unsuppress commands.

Because we entered the values of E_a and A for Equation (11.27) rather than a value for the rate constant itself, we can use the same script for other temperatures simply

---

[2]This example is modified from one in the GWB User's Guide (Bethke, 1994). Program REACT in The Geochemist's Workbench™ also implements several other equations for rate laws, catalyzing and inhibiting species, nucleation, catalysis, and so on.

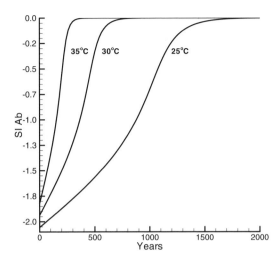

Figure 11.2. The Saturation Index of albite as a function of time at various temperatures, as calculated by REACT.

by changing the statement $T = 25$ to $T = 30$ or $T = 35$, etc. The results of this calculation are shown in Figure 11.2.

It is not difficult to see that although we have successfully modeled a kinetic reaction, there are complicating factors to be considered before we could model a natural or realistic situation. For example, the rate constants for many minerals such as albite are strongly pH dependent. Also, the rates of some reactions will affect the rates of others, because the same solute species may be involved. In a sense, everything that happens affects everything else.

PHREEQC   To illustrate how PHREEQC handles kinetics, we will consider the same reactions. In PHREEQC the user must program the rate equation in a series of statements in a subset of the BASIC language, an option that gives great flexibility, now also available in REACT. For example, to calculate the Saturation Index of albite dissolving in the same solution as shown in Table 11.1, the PHREEQC input script would be similar to that in Table 11.2.

There are two keywords in this script which are unique to kinetic calculations – KINETICS and RATES.

KINETICS   This keyword begins a block of data which identifies kinetic reactions and supplies parameters used in each reaction, such as activation energies, duration of reactions, number of steps, etc. In Table 11.2 only one kinetic reaction is specified, for albite dissolution. For comparison with the REACT example, we use the same amount of mineral, and the same kinetic parameters. The current amount (moles) of mineral, which will decrease during dissolution, is identified as -m, and the initial amount of mineral, in this case the same quantity, is identified as -m0. PHREEQC measures

```
TITLE Albite dissolution kinetics
SOLUTION 1
        temp        25
        units       mol/kgw
        ph          6
        Na          0.3
        Ca          0.05
        Cl          . 3         charge
        C(4)        0.02
EQUILIBRIUM_PHASES 1
        Quartz          0.0
        Kaolinite       0.0
        Muscovite       0.0
KINETICS 1 Define Albite parameters
        Albite
                -m          0.03814  # 10 grams albite
                -m0         0.03814
                -parms      18370  117200
        -steps      6311385e4 in 100 steps
INCREMENTAL_REACTIONS true
RATES
        Albite
        -start
        1         rem M = current no. moles of Ab
        2         rem at start, M (= m) = M0 = .03814
        3         rem Initial area (A0) is 1000 [cm^2/g] for 10 g (=1e4 cm^2)
        4         rem so A0/V [cm^-1] is 10
        5         rem pre-exp const A [mol/cm^2.s] is 18370 = PARM(1)
        6         rem act energy Ea [J/mol] is 117200 = PARM(2)
        7         rem this gives rate const of 5.50e-17 mol/[cm^2.s] at 25C
        10        A0 = 1e4
        15        V = 1000
        20        sr_ab = SR("Albite")
        25        if (M <= 0) then goto 200
        30        R = 8.31451
        35        T = 298.15
        45        area = (M/M0)^(2/3)*A0
        50        rate_const = PARM(1)*EXP(-PARM(2)/(R*T))
        55        rate = (area/V)*rate_const*(1-sr_ab)
        60        rem multiply by 1000 to change mol/(cm^3.s) to mol/(L.s)
        65        rate = rate*1000
        70        moles = rate*TIME
        75        PUT(area,1)
        80        PUT(rate,2)
        200        SAVE moles
        -end
SELECTED_OUTPUT
        -file       albite.sel
        -reset      false
USER_PUNCH
        -headings   Ab_left  area  rate  Time(years)  a_Na+  SI(Albite)
        -start
        10  PUNCH  M  GET(1)  GET(2)  TOTAL_TIME/31556925  10^(LA("Na+"))  SI("Albite")
        -end
END
```

Table 11.2. A PHREEQC script for the kinetic dissolution of albite.

time in seconds, so a duration of 2000 years is quite a large number. The keyword
INCREMENTAL_REACTIONS is made true so that each reaction step begins where the
last one ended. The default, false, means that each step begins at time zero.

RATES   This data block supplies a program written in BASIC which defines the rate
reactions, using parameters specified in the KINETICS block. In addition to most of
the usual BASIC commands, several commands unique to PHREEQC are also available,
which allow access to parameters calculated by PHREEQC. Using a BASIC program means
a bit more work for the user, but the results are worth it, in the sense that virtually any
kinetic rate law may be introduced.

In the example in Table 11.2,

- two parameters are imported from KINETICS, PARM(1) and PARM(2),

- a command special to PHREEQC, SR(''ALBITE''), gives the saturation ratio
  Q/K.

- the initial surface area, A0, is 1000 cm^2, but this decreases as albite dissolves. It
  is assumed here that albite occurs as uniform spheres or cubes, so that the surface
  area decreases as the 2/3 power of the decrease in mass. As noted before, this is
  one of the most difficult parameters to quantify in kinetic studies.

- another keyword not previously mentioned is USER_PUNCH, which is much like
  SELECTED_OUTPUT, but which allows the user to define his/her own output
  variables. If these occur in a BASIC program, they must be saved in an array
  with PUT statements, and accessed in USER_PUNCH with GET statements.

The results from PHREEQC look much like those in Figure 11.2, with some differences
in the Saturation Indices, because of the different data used. A more complex example of
albite dissolution based on the work of Sverdrup (1990) involving temperature effects,
CO_2 concentration, and organic decomposition is included in the database supplied with
PHREEQC.[3]

11.4   Kinetics of Acetate Decomposition

An example of the use of the rate constant and activation energy in a homogeneous reaction is the work of Palmer and Drummond (1986) on the kinetics of the decomposition
of acetate compounds such as acetic acid and sodium acetate at elevated temperatures.
Organic compounds are generally quite unstable at elevated temperatures, but some
persist for surprisingly long times. Just how long is an important geological question,
because these compounds may be involved in many important processes, such as the
generation of natural gas and transport of base metals. To find out how long acetates
could be expected to persist at various temperatures, Palmer and Drummond measured
the concentration of acetate remaining after various times at several temperatures, using

---

[3]A recent release of The Geochemist's Workbench™ adds the ability to program any rate law, much as in
this PHREEQC example.

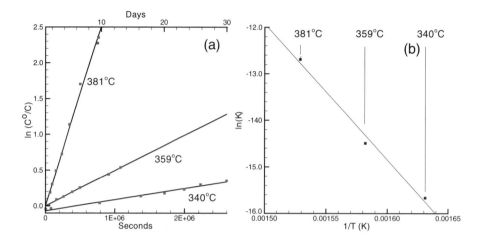

Figure 11.3. (a) The results of experiments by Palmer and Drummond (1986) on the decomposition of sodium acetate at three temperatures. (b) Rate constants derived from the slopes of the lines in (a), in an Arrhenius plot.

several different containing surfaces, because it is found that surfaces play a catalytic role. That is, the rate of reaction depends on the surface available to the reaction. Some of their data are shown in Table 11.3 and Figure 11.3.

The experimental data in Figure 11.3(a) are plotted according to Equation (11.16), so that the slope of each line gives the rate constant at that temperature. These constants are listed in Table 11.4, and plotted in Figure 11.3(b). Assuming that this Arrhenius relationship continues to hold at other temperatures (which implies that the reaction mechanism remains the same at these other temperatures), we can now calculate the rate constant at other temperatures, and therefore calculate how much acetate will remain after various lengths of time. Some typical results are shown in Table 11.5. The rate constant at 200°C from Figure 11.3(b) is 1.0734×10^{-13} s^{-1}. Using this in Equation (11.16) we find that there will still be appreciable acetate concentrations after thousands of years.

Acetate Example Using PHREEQC   The acetate calculation can also be performed using PHREEQC, using a simple BASIC program in the RATES data block as before. The script would be similar to that in Table 11.6.

Part of almost all homogeneous kinetic calculations will be some method to *decouple* the reactive species, which are often redox species. In kinetic calculations, the species are obviously not at equilibrium with each other, at least at the start of the calculation; they approach equilibrium during the calculation. But speciation programs such as PHREEQC and REACT assume all species to be at equilibrium unless told otherwise. In this case we want CH$_4$ to not react with other species, partly to see what it is doing during the reaction, and partly because in nature it is extremely unreactive. We also want acetate to be decoupled because it is metastable and will not even exist in the solution at equilibrium. At the time of writing, this is not necessary in PHREEQC, because the

Time			
hours	seconds	Acetate molality	$\ln(C^o/C)$
Experiment 26, 340°C			
22.0	79 200	0.04E+010	−0.036
219.0	788 400	9.59E−01	0.042
382.5	1 377 000	8.71E−01	0.138
475.5	1 711 800	8.37E−01	0.178
554.5	1 996 200	7.92E−01	0.233
618.0	2 224 800	7.42E−01	0.298
721.0	2 595 600	7.03E−01	0.352
Experiment 27, 359°C			
14.0	50 400	9.89E−01	0.011
43.5	156 600	9.15E−01	0.089
73.0	262 800	8.78E−01	0.130
109.5	394 200	8.22E−01	0.196
140.5	505 800	7.75E−01	0.255
253.5	912 600	6.43E−01	0.442
301.0	1 083 600	5.84E−01	0.538
Experiment 28, 381°C			
1.0	3 600	1.05E+00	−0.044
18.5	66 600	8.28E−01	0.189
26.5	95 400	7.38E−01	0.304
41.5	149 400	6.11E−01	0.493
66.0	237 600	4.84E−01	0.726
98.0	352 800	3.21E−01	1.136
142.0	511 200	1.82E−01	1.704
211.0	759 600	1.03E−01	2.273
213.5	768 600	9.50E−02	2.354

Table 11.3. Data from Palmer and Drummond (1986) on the breakdown of sodium acetate on titanium surfaces at three temperatures.

Expmt	T(°C)	Rate const. (s^{-1})	1/T (K)	$\ln K$
26	340	1.570E−07	0.001631	−15.764
27	359	4.937E−07	0.001582	−14.332
28	381	3.087E−06	0.001529	−12.778

Table 11.4. Rate constants derived from the slopes of the curves in Figure 11.3.

t (years)	t (seconds)	$\ln(C^o/C)$	C (molal)
10	3.156E+08	3.3874E−05	0.099997
100	3.156E+09	3.3874E−04	0.099966
1 000	3.156E+10	3.3874E−03	0.099662
10 000	3.156E+11	3.3874E−02	0.096669
100 000	3.156E+12	3.3874E−01	0.071267
500 000	1.578E+13	1.6937E+00	0.018384
1 000 000	3.156E+13	3.3874E+00	0.003380

Table 11.5. The concentration of sodium acetate remaining after various lengths of time at 200°C, starting with 0.1 molal.

database does not contain a redox reaction for acetate, so while it will ionize and react with metals, forming acetate complexes, it will not decompose. This will undoubtedly be changed in future versions.

In PHREEQC, decoupling is achieved by defining new species and, if desired, their reactions with other species. In this case we simply define Methane to be a new species, and give it no reactions; it is inert. In REACT (below), decoupling is achieved with a decouple statement.

The reaction rate is calculated (in statement 40) according to Equation (11.14), and we increase the duration to 10^6 years. The activation energy E_a (PARM(1)) and the pre-exponential constant A (PARM(2)) are obtained from the linear fit coefficients of Figure 11.3(b); see Equation (11.26). The results of both the PHREEQC output and Equation (11.16) (Table 11.5) are shown in Figure 11.4. They are identical.

Acetate Example Using REACT In REACT, kinetic information is provided in the kinetic statement, as mentioned above. This time, we use the rxn and rate_law parts of the kinetic statement to enter the reaction and the rate equation explicitly. REACT parses the rxn statement to assess what the reaction is, and uses rate_law as the rate equation. The script is shown in Table 11.7.[4] The results are shown, along with those for PHREEQC, in Figure 11.4.

Well, one could say, if the calculation can be done on a spreadsheet, or even a calculator, why should we use a complicated program? It depends on how much we want to know. If we only want to know the amount of acetate as a function of time, a calculator is quite sufficient. However, the PHREEQC and REACT results include data on many solution species besides acetate, the solution pH, Saturation Indices, oxidation states, and so on, which would be of interest in many situations. We could also calculate the effects of adding other components, although this would be subject to experimental confirmation.

Acetate Example Using CKS At this point a mention of a different kind of program seems appropriate. CKS (Chemical Kinetics Simulator) is a program distributed

---

[4]Note that in rate_law we include the molality of each species individually in order to get the total acetate concentration. In a later release, this is accomplished with a single term.

```
DATABASE llnl.dat
TITLE Kinetic decomposition of sodium acetate
definition of Methane decouples CO2 and CH4

SOLUTION_MASTER_SPECIES
    Methane Methane 0.0 CH4 16.0432
SOLUTION_SPECIES
    Methane = Methane
    log_k = 0
SOLUTION 1
    temp    200
    units   mol/kgw
    ph 7
    Na  .1
    Acetate .1
    Cl  .001    charge
KINETICS 1 Define Acetate parameters
    NaAc
#    reaction is NaCH3COO(aq)  + H+  -> Na+  + CH4(aq)  + CO2(aq)
#      or in phreeqc NaAcetate(aq) + H+  -> Na+  + Methane  + HCO3-
        -formula  Acetate -1 H -1 Methane 1 C 1  O 2
        -m      .1
        -m0     .1
        -parms    242908  7.0244e13
# these parameters give a rate constant at 200 C = 1.0734e-13.
    -steps   3.1556925e13 in 100 steps
INCREMENTAL_REACTIONS true
RATES
    NaAc
    -start
    1    rem PARM(1) is Ea
    2    rem PARM(2) is A
    10   R = 8.31451
    20   T = 473.15
    30   rate_const = PARM(2)*EXP(-PARM(1)/(R*T))
    40   rate = rate_const*TOT("Acetate")
    50   moles = rate*TIME
    60   PUT(rate,1)
    200 SAVE moles
    -end
SELECTED_OUTPUT
    -file   acetate.sel
    -reset  false
USER_PUNCH
    -headings    Time(yrs) rate  Acetate C(4) Methane
    -start
    10  PUNCH  TOTAL_TIME/31556925 GET(1) TOT("Acetate") TOT("C(4)") TOT("Methane")
    -end
END
```

Table 11.6. A PHREEQC script for the kinetic decomposition of sodium acetate at 200°C.

```
time start = 0 days, end = 1e6 years
temperature = 200
decouple Acetic\_acid(aq)
decouple Methane(aq)
swap CO2(aq) for HCO3-
1 kg free H2O
total molality CO2(aq) = 1e-10
total molality Methane(aq) = 1e-10
total molality Na+ = .1
total molality Acetic\_acid(aq) = .1
total molality Cl- = .001
pH = 7
kinetic redox-1 \
rxn = "Acetate  + H+   -> Methane(aq)   + CO2(aq)" \
rate\_con = 1.073e-13 \
rate\_law = 'rate\_con * (molality("Acetate") \
+ molality("NaCH3COO(aq)") + molality("Na(CH3COO)2-") \
+ molality("Acetic\_acid(aq)"))'
```

Table 11.7. A REACT script for the kinetic decomposition of sodium acetate at 200°C.

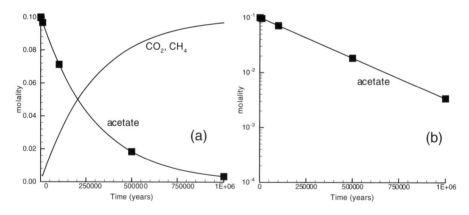

Figure 11.4. (a) Concentration of sodium acetate remaining as a function of time at 200°C, using data of Palmer and Dummond (1986). The solid line is calculated by PHREEQC, by REACT, and by program CKS, which all give virtually the same results. The square points are from Table 11.5. (b) The same data using a log scale for the acetate concentrations.

free by IBM (http://www.almaden.ibm.com/st/msim/) that models kinetic reactions by considering a large number of particles and a reaction or series of reactions between "species", which are simply labeled particles. Changes in the system are modeled by randomly selecting among probability-related reaction steps, where the probabilities are related to the reaction rate constants. It is ideally suited to trying out various reaction schemes in complex systems, and comparing the results with experimental data. In our acetate decomposition example, the results are virually indistinguishable from those of PHREEQC and REACT. However, it is more of a research tool than an environmental modeling tool, and will not be discussed further.

11.5 Coupled Aqueous Speciation and Biological Processes

A good example of interactions among inorganic chemistry, organic compounds degradation, and microbial activities is formation and destruction of metal–chelate complexes. At many defense facilities in the USA, radionuclides and chelating agents are buried together (mixed wastes). The formation of radionuclide–chelate complexes greatly increases radionuclide mobility and poses significantly more risks to humans and the environment. Microbial degradation of chelating agents has the apparent benefits of reducing the mobility and risk.

However, complex interactions or feedback loops exist among the different components of the system:

- bacteria can use the chelate as the substrate, and thus accomplish the goal of reducing the mobility of radionuclides by increase of sorption and less solubility;

- on the other hand, biodegradation of organic compounds builds up the bicarbonate content of the system, and forms mobile actinide–carbonate complexes;

- actinides when present at micro-molar per liter levels can be toxic to microorganisms and hence inhibit microbial growth;

- microbiological degradation reactions cause changes of pH and hence change of the speciation, sorption, and solubility of actinides;

- most actinides have multiple valence states, and microbial activities can change the redox states of the radionuclides and change their mobility.

With all these complex interactions, it is desirable to build a model that can quantitatively evaluate all the components of the system. Rittmann and VanBriesen (1996) and VanBriesen and Rittmann (1999) developed the code CCBATCH to do just that. CCBATCH, which stands for "co-contaminant batch reactor", has three components.

Aqueous speciation   This is in line with traditional speciation code like MINTEQA2 and PHREEQC. Instantaneous equilibrium is assumed for all homogenous aqueous complexation reactions.

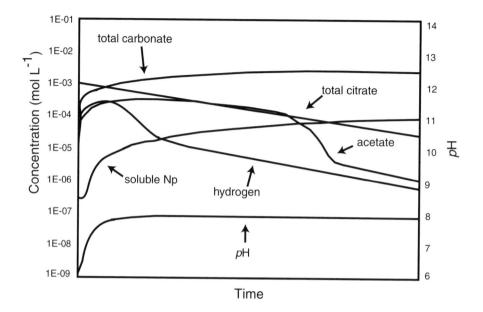

Figure 11.5. Simulated fate of Np during the mineralization of citrate by an anaerobic microbial consortium. After Banaszak *et al.* (1998).

Biodegradation   Degradation of organic compounds is treated as a kinetic process and described by the dual Monod kinetics formulation (Bae and Rittmann, 1996).

Cell growth and decay   The growth and decay of the biomass (cells) are also simulated as a mass balance.

CCBTACH solves the aqueous speciation equations and the Monod kinetics sequentially. Biodegradation reactions are solved in the biodegradation subroutine or submodel. The products from biodegradation reactions are brought into the speciation subroutine, where the speciation is changed accordingly due to the changes of, for example, total H^+ and CO_3^{2-}.

An example calculation was given in Banaszak *et al.* (1998) for anaerobic degradation of citric–Np complexes by a consortium of bacteria. The initial conditions are pH 6, total carbonate concentration of 10^{-5} M, fixed redox potential of -250 mV, and equilibrium with amorphous $Np(OH)_4$. The modeling results are shown in Figure 11.5. The initial degradation of citrate results in an increase of pH and a slight decrease of Np in solution. However, degradation of citrate and acetate produces carbonate, and that leads to the formation of carbon–Np complexes and an increase of 2–3 orders of magnitude in total dissolved Np concentrations. The results are somewhat counter-intuitive – degradation of organic chelate in a reducing environment may not necessarily reduce actinide mobility. These modeling results are difficult to verify with laboratory and field data. Nevertheless, it illustrates the importance of considering the interplay of various processes, and the complexity in this relatively simple system.

$Mn^{2+} + 0.5O_2 + H_2O \rightarrow MnO_2 + 2H^+$	1
$Fe^{2+} + 0.25O_2 + 2.5H_2O \rightarrow Fe(OH)_3 + 2H^+$	2
$2Fe^{2+} + MnO_2 + 4H_2O \rightarrow 2Fe(OH)_3 + Mn^{2+} + 2H^+$	3
$NH_4^+ + 2O_2 \rightarrow NO_3^- + 2H^+ + H_2O$	4
$H_2S + 2O_2 \rightarrow SO_4^{2-} + 2H^+$	5
$H_2S + MnO_2 + 2H^+ \rightarrow Mn^{2+} + S^o + 2H_2O$	6
$H_2S + 2Fe(OH)_3 + 4H^+ \rightarrow 2Fe^{2+} + S^o + 6H_2O$	7
$FeS + 2O_2 \rightarrow Fe^{2+} + SO_4^{2-}$	8
$CH_4 + 2O_2 \rightarrow CO_2 + 2H_2O$	9
$CH_4 + SO_4^{2-} + 2H^+ \rightarrow H_2S + CO_2 + 2H_2O$	10

Table 11.8. Secondary redox reactions in the model presented by Hunter *et al.* (1998).

The Geochemist's Workbench™ program REACT now includes procedures to calculate reaction kinetics of biotransformations, and the PHREEQC documentation gives an example of similar calculations in PHREEQC.

11.6   Application to Landfill Leachate into Aquifers

Municipal landfill leachate usually contains high concentrations of dissolved organic carbon (DOC), and hence intrusion of leachate into aquifers can result in development of redox zones due to microbial activities. Redox zones developed in oxic aquifers show a sequence, from landfill to downgradient areas, methanogenic/sulfate-reducing, iron-reducing, manganese-reducing, nitrate-reducing, to aerobic zones (Amirbrahman *et al.*, 1998; Baedecker and Back, 1979; Baedecker *et al.*, 1993; Bjerg *et al.*, 1995). These microbiallly mediated redox reactions have significant influence on inorganic pore fluid chemistry, natural attenuation, and fate and transport of toxic elements and compounds in the leachate.

Hunter *et al.* (1998) used a kinetic model to simulate reactive transport and groundwater evolution upon the intrusion of landfill leachate into an oxic aquifer. Most interestingly, they found that the patterns of redox fronts are similar to what have been seen in marine sediments. Hunter *et al.* (1998) first developed a one-dimensional multicomponent reactive transport model BIORXNTRN and then conducted a numerical experiment using this code. The actual model is quite complex; we describe here the essentials of it so that we can see what can be done.

In the model, an anoxic leachate from a landfill has infiltrated into an oxic aquifer for 20 years. The flow velocity is 10 m yr^{-1}, and longitudinal dispersivity is 4 m. The leachate is rich in organic carbon. Degradation of organic carbon via microbial activities causes drastic changes in pore fluid inorganic chemistry and mineralogy. Several zones are developed (Figure 11.6).

Within the first 20 m, all dissolved and solid oxidants are depleted (Figure 11.6b,c). Microbial metabolism uses DOC and electron acceptors in the sequence O_2, NO_3^-, SO_4^{2-}, then methanogenesis becomes the dominant pathway for biodegradation. When

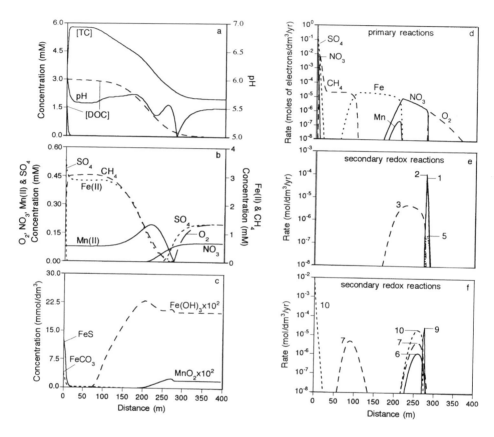

Figure 11.6. Chemical zonation in an aquifer after 20 years leachate contamination. Groundwater flows from left to right. (a) and (b) Dissolved species distribution; (c) Solid species; (d) Primary pathways for DOC biodegradation. (e) and (f) Rates of secondary reactions listed in Table 11.8. After Hunter *et al.* (1998).

DOC is depleted at about 280 m, a reverse zonation occurs before it returns back to aerobic conditions of the uncontaminated aquifer.

Accompanying the biodegradation, pH drops about one unit from 6 to 5. $Fe(OH)_3$ and MnO_2 are reduced and dissolved. The pore fluids begin to be supersaturated with $FeCO_3$ and $FeCO_3$ begins to precipitate. Reduction of SO_4^{2-} also produces H_2S and causes super-saturation and precipitation of FeS. These patterns of change have been observed in the field and compare well with marine sediment diagenesis (Stumm and Morgan, 1996).

We should note how Hunter *et al.* (1998) treat the kinetics in their model. They first divided reactions into different types.

- *Primary Redox Reactions.* Hunter *et al.* referred to the reactions of the redox sensitive species with DOC via microbial activities. A first-order rate law with respect to organic carbon concentration is used for these reactions. Two fractions of organic matter with different degradation rates are used.

- *Secondary Redox Reactions.* These are the redox reactions that involve products of the primary redox reactions and are essentially abiotic. They use a second-order, bimolecular rate expressions for these reactions.

- *Precipitation and Dissolution Reactions.* The apparent rate constants are a function of saturation state.

- *Aqueous speciation, Acid–base Reactions, and Sorption of* NH_4^+. These reactions are treated as being at equilibrium.

These modeling results show complex coupling and feedback mechanisms between biogeochemical reactions and inorganic chemistry. Hunter *et al.* (1998) relate this to a reaction network. Unlike the local equilibrium models we described in the preceding chapters, kinetically controlled reactions are pathway dependent and there may exist multiple pathways. Our understanding of these systems is limited. One difficulty for practical applications of such models is determination of a realistic rate constant. Hunter *et al.* (1998) noted that the *in situ* organic degradation rates reported in the literature span over ten orders of magnitude, which makes field application of this kind of kinetic model difficult. They conducted numerical experiments using various rates, and the results are quite different.

11.7   Conclusions

We have already stated (§1.4.2 and §2.5) that the subject of chemical reaction kinetics is both important and poorly understood in geochemical modeling. The problem is that this subject is much more difficult than equilibrium thermodynamics. The kinetics of a few reactions are fairly well understood, but a great many more have completely unknown kinetics. Furthermore, those that are understood have usually been studied in fairly "clean" systems, and the effect of added components is uncertain. Even if some reactions were perfectly understood, the usefulness of including the kinetics of only some of the large number of reactions present in any realistic modeling study is questionable.

One way of getting around this is to use what seem to be reasonable estimates of rate laws and rate constants, as in some of our examples. This has the very useful effect of showing the importance of including kinetics in modeling studies, but the absence of reliable data for each reaction certainly renders the results of questionable usefulness.

It is safe to say that, at the present time, no realistic modeling study (that is, one based on a natural situation) has been carried out with realistic kinetics for all reactions involved. This is one of the factors used in evaluating the results of modeling studies, whether based entirely on equilibrium theory or whether they use some kinetics. There is a great deal of work to be done in this challenging field.

Appendix A

Modifying a Database

A.1   Why Modify a Database?

We have already mentioned that although most modeling programs come with an associated database, the responsibility for using appropriate data is always that of the user, not the programmer or agency which supplies the database. There are basically two situations which might call for the user to change or add to the data supplied with the program.

- One is, of course, if the data become obsolete; i.e., new or better data become available, superseding those already in the database. Because most users are not solution chemistry experts, and/or have no time to keep up with the latest developments in solution and surface chemistry, this may not be a common situation.

- It is not uncommon, however, to find that the database simply has no data for some element that becomes important to us. We must then obviously find some data, and add them to the database. The easiest way to do this is simply to take data from the database of some other modeling program that happens to be more complete with respect to the missing element or elements. Alternatively, we can consult the primary literature.

In either case, it becomes important to understand the structure of the databases involved, because the modeling program expects to find the data in a certain format.[1] To understand and conform to this format, we must read the program documentation, and in doing this it helps to understand the fundamentals of thermodynamics as given in Chapter 3. In this book we cannot discuss the structure of all databases that may be encountered, but it may be helpful to go through one example in detail.

---

[1]An exception is MINTEQA2, in which we alter or add data by an interactive question-and-answer process using PRODEFA2. Whether this tedious process is an improvement is an open question.

253

A.1.1   Adding Arsenic Data to PHREEQC

If we are using PHREEQC, and we have a problem involving arsenic, we soon find that the PHREEQC database contains no data for this element.[2] We can either give up, change to another program, or add arsenic data to the PHREEQC database. As PHREEQC is one of the most commonly used geochemical modeling programs, and perhaps the only one we are quite familiar with, we may be stuck with the third option.

Finding Arsenic Data

As mentioned, the simplest way to find required data is to rob a more complete database. Several such datasets have been prepared by members of the Lawrence Livermore National Laboratories (LLNL) geochemical modeling group, formatted so as to be directly readable by The Geochemist's Workbench™ group of programs, EQ3/6, or SUPCRT92.[3] Note that being "more complete" makes no implications about the accuracy or internal consistency of the data. It just means there are a lot of data.

Database Format Examples

To extract all relevant data for arsenic, we could search the databases for "As", to find all relevant entries.

The GWB Database   As an example, we use the GWB database, as supplied by LLNL. Normally, all the basis species are listed first. In GWB, the basis species for arsenic is $As(OH)_4^-$, the entry for which is

```
As(OH)4-
     charge= -1.0        ion size=  4.0 A       mole wt.=  142.9508
     3 elements in species
        4.000 O                 4.000 H                1.000 As
```

Other (secondary) species of arsenic are then listed farther down in the database. An example is $As(OH)_3$, which is related to the basis species by the reaction (shown in the database entry)

$$As(OH)_3 + H_2O = As(OH)_4^- + H^+$$

```
As(OH)3
     charge=  0.0        ion size=  4.0 A       mole wt.=  125.9435 g
     3 species in reaction
      -1.000 H2O                 1.000 As(OH)4-       1.000 H+
        -9.6882    -9.2327    -8.7596    -8.4343
        -8.2822    -8.3962   500.0000   500.0000
```

The equilibrium constant logarithms at various temperatures (see §4.3.1) are given; that for 25°C is -9.2327.

---

[2]This Appendix was written before database lln1.dat became available for PHREEQC, which does contain arsenic data. However, the comments apply to any data we may wish to add.

[3]Available from the LLNL anonymous ftp site. ftp to "s122.es.llnl.gov", sign on as "anonymous", and give your e-mail address when prompted for a password. The datasets are located in directories "/users/johnson/ ... ", where ... refers to /gwb, /eq36, or /supcrt92.

The EQ3/EQ6 database   For contrast, compare the format of the EQ3/EQ6 database for
the same complex:

```
As(OH)3(aq)
      sp.type =  aqueous
*     EQ3/6   =  com, alt, nea
      revised = 12-apr-1990
*     mol.wt. = 125.944 g/mol
*     DHazero =  3.0
      charge  =  0.0
****
      3 element(s):
       1.0000 As             3.0000 H             3.0000 O
****
      3 species in aqueous dissociation reaction:
     -1.0000  As(OH)3(aq)                   1.0000  H+
      1.0000  H2AsO3-
*
**** logK grid [0-25-60-100C @1.0132bar; 150-200-250-300C @Psat-H2O]:
         -9.6789   -9.2048   -8.7528   -8.4285
         -8.2777   -8.3928  500.0000  500.0000
*
*     gflag = 2 [calculated delG0f(delH0f,S0PrTr) used]
*     extrapolation algorithm: 69hel
*     ref-state data   [source: 92gre/fug ]
*         delG0f =   -639.681 kj/mol       [reported]
*         delG0f =   -639.682 kj/mol       [calculated]
*         delH0f =   -742.200 kj/mol       [reported]
*         S0PrTr =    195.000 j/(mol*K)    [reported]
```

Note that $As(OH)_3$ is defined in terms of a different basis species ($H_2AsO_3^-$), so
that equilibrium constants cannot be directly compared between databases. They can
of course be recalculated quite easily. In the EQ3/EQ6 database, we are also supplied
with thermodynamic data about the species which is not directly used by the program,
but which might be useful.

Now we have the data for an arsenic species, and the problem is to get it into
PHREEQC. At this point, we have two choices:

1. we can add data to PHREEQC.DAT, the PHREEQC database, or

2. we can add data to individual input files to PHREEQC, which will be treated just
   as if they were obtained from the database.

It is probably always best to begin by adding data to input files, rather than to the
database, until we are quite sure of what we are doing, especially if we are not the
only user of the database. PHREEQC allows us to do this quite easily, using the special
input file keywords SOLUTION_MASTER_SPECIES, SOLUTION_SPECIES, and
SURFACE_SPECIES.

Aqueous Species   The keyword SOLUTION_MASTER_SPECIES allows us to define
new basis species. To see how this is done, the easiest thing is to look into PHREEQC.DAT
to see how others are defined. Since it has not previously been defined, we have a choice

of As species to use a basis species, but of course in this case the simplest thing is to use the same species as the database we are borrowing from, i.e., $As(OH)_4^-$. We find that in PHREEQC.DAT, basis species are defined in the format

```
#
#element  species        alk     gfw_formula     element_gfw
#
As        As(OH)4-       0.0     As              74.9216
```

which includes an alkalinity factor, the formula to be given an atomic weight, and the atomic weight. The usage of these terms is explained in the documentation, but we don't really need to fully understand all the details – we can usually figure out what we need to do by looking at the other entries.

Other aqueous arsenic species are then entered using the SOLUTION_SPECIES keyword. Again copy the format given in PHREEQC.DAT. For $As(OH)_3$, this is

```
As(OH)4- + H+ = H3AsO3 + H2O
      log_k       9.2327
```

It is important to note that in PHREEQC, secondary species are always defined in reactions in which the new species immediately *follows* the = sign, whereas in GWB, the secondary species is the first term on the left side of the equation. Therefore the log K values have opposite signs in the two databases.

Surface Species   The keyword SURFACE_SPECIES allows us to define new surface species in the input file. Unfortunately, it is a little more difficult to rob another database in this case, because those available from LLNL contain no surface data. The most common source of these data is Dzombak and Morel (1990). A number of surface reactions between arsenic and HFO are available in various tables in this reference.

For example, in Table 10.7 of Dzombak and Morel we find the surface complexation reaction

$$\equiv FeOH^\circ + H_3AsO_3^\circ = \equiv FeH_2AsO_3^\circ + H_2O; \quad \log K_1 = 5.41$$

By observing how surface reactions are entered into PHREEQC.DAT (which is the same as for aqueous reactions, but using the surface site species noted in Table 7.1), we translate this into a PHREEQC reaction:

```
Hfo_wOH + H3AsO3 = Hfo_wH2AsO3 + H2O
log_k  5.41
```

and enter it under the keyword SURFACE_SPECIES in the input file. Similarly for other arsenic reactions from Dzombak and Morel (1990), or from other sources in the literature.

The Input File

Finally, we are ready to prepare an input file, using arsenic data. Following an example used by Bethke (1996, Chap. 12), we calculate the adsorption of As and Zn (which is already in PHREEQC.DAT) on HFO in a simple NaCl solution, from *p*H 4 to *p*H 12. The input file is

```
TITLE Simulation of As and Zn adsorption on HFO.
Aqueous As data obtained from LLNL, and surface data from D&M (1990).
(phreeqc.dat contains no data for As).
Oxidizing conditions (pe = 12).

SOLUTION_MASTER_SPECIES
#
#element species        alk     gfw_formula      element_gfw
#
As      As(OH)4-        0.0     As               74.9216
SOLUTION_SPECIES

As(OH)4- = As(OH)4-
        log_k       0.000
As(OH)4- +0.5O2 = AsO4-3 + H2O + 2H+
        log_k       12.1848
As(OH)4- +H+ = H3AsO3 +H2O
        log_k       9.2327
As(OH)4- = AsO2OH-2 + H2O + H+
        log_k       -11.0123
As(OH)4- + 0.5O2 = HAsO4-2 + H+ + H2O
        log_k       23.7806
As(OH)4- + 0.5O2 = H2AsO4- + H2O
        log_k       30.5387
As(OH)4- + 0.5O2 + H+ = H3AsO4 + H2O
        log_k       32.7890
As(OH)4- + 4H+ + 2SO4-2 = AsS2- + 4H2O + 4O2
        log_k       -249.8833
As(OH)4- + H+ = AsH3 + H2O + 1.5O2
        log_k       -124.6488
As(OH)4- + 5H+ + 2SO4-2 = HAsS2 + 4H2O + 4O2
        log_k       -246.1818
SURFACE_SPECIES

    Hfo_wOH + H3AsO3 = Hfo_wH2AsO3 + H2O
    log_k  5.41

    Hfo_wOH + AsO4-3 + 3H+ = Hfo_wH2AsO4 + H2O
    log_k  29.31

    Hfo_wOH + AsO4-3 + 2H+ = Hfo_wHAsO4- + H2O
    log_k  23.51

    Hfo_wOH + AsO4-3 = Hfo_wOHAsO4-3
    log_k  10.58
SURFACE 1
        Hfo_sOH        5e-6     600.     0.09
        Hfo_wOH        2e-4
SOLUTION 1
    -units mmol/kgw
    pH  4.0
    pe  12
    As  .1
    Zn  .1
    Na  100
    Cl  100 charge
USE surface none
#
# Model definitions
#
PHASES
        Fix_H+
        H+ = H+
        log_k  0.0
END
SELECTED_OUTPUT
        -file surf_12.pun
        -reset false
        -ph    true
        -totals Zn As
USE solution 1
USE surface 1
EQUILIBRIUM_PHASES 1
        Fix_H+   -4.0    NaOH    10.0
END
```

```
USE solution 1
USE surface 1
EQUILIBRIUM_PHASES 1
        Fix_H+    -4.25   NaOH     10.0
END
USE solution 1
USE surface 1
EQUILIBRIUM_PHASES 1
        Fix_H+    -4.5    NaOH     10.0
END
USE solution 1
USE surface 1
EQUILIBRIUM_PHASES 1
        Fix_H+    -4.75   NaOH     10.0

........................etc.

USE solution 1
USE surface 1
EQUILIBRIUM_PHASES 1
        Fix_H+    -12.0   NaOH     10.0
END
```

Note that we have asked for the pH and total aqueous concentrations of As and Zn to be printed out in a special file called `surf_12.pun`, and suppressed all default output with `-reset false`. By transferring this file to a spreadsheet, subtracting the As and Zn concentrations from the initial concentrations, and dividing by the initial concentrations, we obtain the fraction of the initial concentration that has been adsorbed onto the HFO.

Alternatively, we can use another keyword, USER_PUNCH, which allows us to use simple BASIC programming statements to do the necessary calculations, putting the results into `surf_12.pun`, and obviating the need for using a spreadsheet. Thus, the following lines could be added after (or before) the SELECTED_OUTPUT statement:

```
USER_PUNCH
        -headings  As_fraction  Zn_fraction
        -start
10    frac_As = ((0.1/1000)-TOT("As"))/(0.1/1000)
20    frac_Zn = ((0.1/1000)-TOT("Zn"))/(0.1/1000)
30    PUNCH  frac_As  frac_Zn
        -end
```

Now `surf_12.pun` will contain the fractions adsorbed, as well as the pH and aqueous concentrations. The use of variables such as TOT(``As''), which enable access to quantities used or calculated internally by PHREEQC, is explained in the documentation.

Comparison of Results from PHREEQC and GWB

To make sure things are working correctly, it is always useful to compare results from different programs. In this case, we used program REACT from The Geochemist's Workbench™ to simulate the same system. The results are shown in Figure A.1. It would seem that the programs are working as expected.

Note that we specified oxidizing conditions ($pe = 12$ at $pH = 4$). Different results would be obtained for reducing conditions, because As exists in nature as As^{3+} under reducing conditions, and as As^{5+} under oxidizing conditions. Different species are therefore involved, which react somewhat differently with the HFO surface. A log f_{O_2}–pH diagram for As is shown in Figure A.2, showing the various oxyhydroxide species.

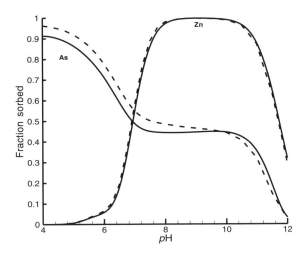

Figure A.1. Fractions of total As and Zn adsorbed onto hydrous ferric oxide (HFO) surface as a function of pH. Solid lines: program REACT (The Geochemist's Workbench™). Dashed lines: PHREEQC.

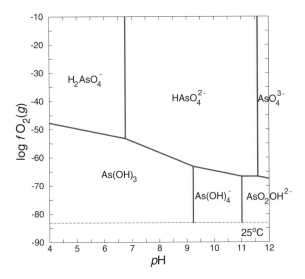

Figure A.2. Aqueous arsenic species as a function of $\log f_{O_2}$ and pH. Total As is 10^{-4} molal.

.

References

Adamson, A.W., and Gast, A.P., 1997. *Physical Chemistry of Surfaces*, 6th edn. New York, John Wiley & Sons, 784 pp.

Allison, J.D., Brown, D.S., and Novo-Gradac, K.J., 1991. MINTEQA2/PRODEFA2, a geochemical assessment model for environmental systems: version 3.0 user's manual. U.S. Environmental Protection Agency Report EPA/600/3–91/021.

Alpers, C.N., Nordstrom, D.K., and Ball, J.W., 1989. Solubility of jarosite solid solutions precipitated from acid mine waters, Iron Mountain, California, U.S.A. Sci. *Geol. Bull.* v. 42, pp. 281–298.

Alpers, C.N., Blowes, D.W., Nordstrom, D.K., and Jambor, J.L., 1994. Secondary minerals and acid mine-water chemistry. Chapter 9, in J.L. Jambor and D.W. Blowes eds, *Handbook on Environmental Geochemistry of Sulfide Mine-wastes*, Waterloo, Ontario, Mineral. Soc. Canada, v. 22, May, pp. 247–270.

American Public Health Association, 1999. Standard methods for the examination of water and wastewater. 20th edn.

Amirbrahman, A., Schonenberger, R., and Johnson, C.A., and Sigg, L, 1998. Aqueous- and solid phase biogeochemistry of a calcareous aquifer system downgradient from a municipal solid waste landfill (Winterthru, Switzerland). *Environ. Sci. & Tech.*, v. 32, pp. 1933–1940.

Anderman, E.R., Hill, M.C., and Poeter, E.P., 1996. Two-dimensional advective transport in ground-water flow parameter estimation. *Ground Water*, v. 34, no. 6, pp. 1101–1109.

Anderson, G.M., 1976. The accuracy and precision of calculated mineral dehydration equilibria, in D.G. Fraser, ed., *Thermodynamics in Geology*, Proceedings of NATO Advanced Study Institute. Oxford, D. Reidel Publishers, pp. 115–136.

Anderson, G.M., 1977. Uncertainties in calculations involving thermodynamic data, in H.J. Greenwood, ed. *Applications of Thermodynamic to Petrology and Ore Deposits*, Vancouver, Mineralogical Association of Canada, pp. 199–215.

Anderson, G.M., 1996. *Thermodynamics of Natural Systems*. New York, John Wiley & Sons. 382 pp.

261

Anderson, G.M., and Crerar, D.A., 1993. *Thermodynamics in Geochemistry – The Equilibrium Model.* New York, Oxford University Press, 588 pp.

Appelo, C.A.J., and Postma, D., 1993. *Geochemistry, Groundwater, and Pollution.* Rotterdam, Brookfield, A.A. Balkema, 536 pp.

Apps, J.A., 1992. Current geochemical models to predict the fate of hazardous wastes in the injection zones of deep disposal wells. Lawrence Berkeley Laboratory report, LBL-26007.

ASTM (American Society of Testing and Material), 1984. Standard practice for evaluating environmental fate models of chemicals. Annual book of ASTM standards. Philadelphia, ASTM, E978-84.

Bae, J.M., and Rittmann, B.E., 1996. A structured model of dual-limitation kinetics. *Biotech. & Bioeng.*, v. 49, pp. 683–689.

Baedecker, M.J., and Back, W., 1979. Modern marine sediments as a natural analog to the chemically stressed environment of a landfill. *J. Hydrology*, v. 43, pp. 393–414.

Baedecker, M.J., Cozzarelli, I.M., Eganhouse, R.P., Siegei, D.I., and Bennett, P.C., 1993. Crude oil in a shallow sand gravel aquifer. III. Biogeochemical reactions and mass balance modeling in anoxic groundwater. *Appl. Geochem.*, v. 8, pp. 569–586.

Bahr, J. M., 1990. Kinetically influenced terms for solute transport affected by heterogenous and homogeneous classical reactions. *Water Resources Res.*, v. 26, pp. 21–34.

Bahr, J. M., and Rubin, J., 1987. Direct comparison of kinetic and local equilibrium formulations for solute transport affected by surface reactions. *Water Resources Res.* v. 23, pp. 438–452.

Banaszak, J.E., VanBriesen, J.M., Rittmann, B.E., and Reed, D.T., 1998. Mathematical modeling of the effects of aerobic and anaerobic chelate biodegradation on actinide speciation. *Radiochimica Acta*, v. 82, pp. 445–451.

Bear, J., 1972. *Dynamics of Fluids in Porous Media.* New York, Dover Publications, Inc., 764 pp.

Bethke, C.M., 1994. The Geochemist's Workbench™, version 2.0, A User's Guide to Rxn, Act2, Tact, React, and Gtplot. Hydrogeology Program, University of Illinois.

Bethke, C.M., 1996. *Geochemical Reaction Modeling.* New York, Oxford University Press, 397 pp.

Bethke, C.M., and Brady, P.V., 2000. How the K_d approach undermines ground water cleanup. *Ground Water*, v. 38, no. 3, pp. 435–443.

Bjerg, P.L., Rugge, K., Pedersen, J.K., and Christensen, T.H., 1995. Distribution of redox-sensitive groundwater quality parameters downgradient of a landfill (Grindsted Denmark). *Environ. Sci. Tech.*, v. 29, pp. 1387–1397.

Bockris, J.O'M., and Reddy, A.K.N., 1970. *Modern Electrochemistry*. New York, Plenum, vol. 1, 622 pp.; vol. 2, 1432 pp.

Boulding, J.R., 1990. Assessing the geochemical fate of deep-well-injected hazardous waste – A reference guide. EPA/625/6-89/025a.

Bredehoeft, J.D., and Konikow, L.F., 1993. Groundwater models: validate or invalidate. *Ground Water*, v. 31, pp. 178–179.

van Breukelen, B.M., Appelo, C.A.J., and Olsthoorn, T.N., 1998. Hydrogeochemical transport modeling of 24 years of Rhine water infiltration in the dunes of the Amsterdam Water Supply. *J. Hydrology*, v. 209, pp. 281–296.

Brothers, K., and Katzer, T., 1987. Artificial recharge to the Las Vegas Valley groundwater system, Clark County, Nevada. Abstracts with Programs – Geological Society of America, vol. 19, no. 7, p. 602.

Brown, J.G., and Eychaner, J.H., 1988. Simulation of five ground-water withdrawal projections for the Black Mesa area, Navajo and Hopi Indian Reservations, Arizona. U.S. Geol. Survey Water-Resources Invest., 88-4000, 51 pp.

Brown, J.G., Bassett, R.L., and Glynn, P.D., 1998. Analysis and simulation of reaction transport of metal contaminants in ground water in Pinal creek basin, Arizona. *J. Hydrology*, v. 209, pp. 225–250.

Bryant, S.L., Schechter, R.S., and Lake, L.W., 1987. Mineral sequences in precipitation/ dissolution waves. *AIChe J.*, v. 33, no. 8, pp. 1271–1287.

Cederberg, G.A., Street, R.L., and Leckie, J.O., 1985. A groundwater mass transport and equilibrium chemistry model for multicomponent systems. *Water Resources Res.*, v. 21, pp. 1095–1104.

Chao, T.T., 1972. Selective dissolution of manganese oxides from soils and sediments with acidified hydroxylamine-HCl. *Soil Sci. Am. J.*, v. 36, pp. 764–768.

Chao, T.T, and Zhou, L., 1983. Extraction techniques for selective dissolution of amorphous iron oxide from soils and sediments. *Soil Sci. Am. J.*, v. 47, pp. 225–232.

Chapelle, F.H., 2000. *Ground-water Microbiology and Geochemistry*. New York, John Wiley & Sons, Inc., 477 pp.

Chapelle, F.H., and Lovley, D. R., 1990. Rates of bacterial metabolism in deep coastal-plain aquifers. *Appl. & Environmental Microbiology*, v. 56, pp. 1865–1874.

Cherry, H. A., Gillham, R.W., and Barker, J.F., 1984. Contaminants in groundwater processes, in *Studies in Geophysics, Groundwater Contaminations*. Washington D.C., National Academy Press, pp. 46–64.

Cooley, M.E., Harshbarger, J.W., Akers, J.P., and Hardt, W.F., 1969. Regional hydrogeology of the Navajo and Hopi Indian Reservations, Arizona, New Mexico, and Utah. U.S. Geol. Survey Prof. Paper 521-A, 61 pp.

Cornell, R.M., and Schwertmann, U., 1996. *The Iron Oxides – Structure, Properties, Reactions, Occurrence and Uses*. New York, Weinheim, 570 pp.

Criscenti, L.J., Laniak, G.F., and Erikson, R.L., 1996. Propagation of uncertainty through geochemical code calculations. *Geochim. Cosmochim. Acta*, v. 60, pp. 3551–3568.

Cullity, B.D., 1978. *Elements of X-ray Diffraction*. Reading, MA, Addison-Wesley Publishing Co, 555 pp.

Davis, A., and Ashenberg, D., 1989. The aqueous geochemistry of the Berkeley Pit, Butte, Montana. *Appl. Geochem.*, v. 4, pp. 23–36.

Davis, J.A. and Kent, D.B., 1990. Surface complexation modeling in aqueous geochemistry, in M.F. Hochella and A.F. White, eds. *Mineral-water interface geochemistry* Review of Mineralogy, v. 23, pp. 177–260. Mineralogical Society of America, Washington, D.C.

Davis, A., Ruby, M. V., Bloom, M., Schoof, R., Freeman, G., and Bergstrom, P. D., 1991. Mineralogic constraints on the bioavailability of arsenic in smelter-impacted soils. *Environ. Sci. & Tech.*, v. 30, pp. 392–399.

Davis, A., Ruby, M.V., and Bergstrom, P.D., 1992. Bioavailability of arsenic and lead in soils from the Butte, Montana, mining district. *Environ. Sci. & Tech.*, v. 26, pp. 461–468.

Davis, A., Drexler, J.W., Ruby, M.V., and Nicholson, A., 1993. Micromineralogy of mine wastes in relation to lead bioavailability, Butte, Montana. *Environ. Sci. & Tech.*, v. 27, pp. 1415–1425.

Davis, A., Ruby, M.V., Bloom, M., Schoof, R., Freeman, G., and Bergstrom, P.D., 1996. Mineralogic constraints on the bioavailability of arsenic in smelter-impacted soils. *Environ. Sci. & Tech.*, v. 30, pp. 392–399.

Davis, J.A., and Leckie, J.O., 1978. Surface ionization and complexation at the oxide/water interface: Surface properties of amorphous iron hydroxide and adsorption of metal ions. *J. Colloids & Interfacial Sci.*, v. 67, pp. 90–107.

Davis, J.A., and Leckie, J.O., 1980, Surface ionization and complexation at the oxide/water interface: Adsorption of anions. *J. Colloids & Interfacial Sci.*, v. 74, pp. 32–43.

Davis, J.A., James, R.O., and Leckie, J.O., 1978, Surface ionization and complexation at the oxide/water interface. I. Computation of electrical double layer properties in simple electrolyte. *J. Colloids & Interfacial Sci.*, v. 63, pp. 480–499.

Davis, J.A., Coston, J.A., Kent, D.B., and Fuller C.C., 1998. Application of the surface complexation concept to complex mineral assemblages. *Environ. Sci. & Tech.* v. 32, pp. 2820–2828.

Davis, S.N., 1988. Where are the rest of analyses? *Ground Water*, v. 26, no.1, pp. 2–5.

Decker, D.L., 1996. The determination of the hydraulic flow and solute transport parameters for several heap leach materials. Unpublished M.S. Thesis, University of Nevada, Reno.

Dixon, D.G., Dix, R.B., and Comba, P.G., 1993. A mathematical model for rinsing of reagents from spent heaps, in J. Hager, B. Hansen, W. Imrie, J. Pusatori and V. Ramachandran, eds, *Extraction and Processing for the Treatment and Minimization of Wastes*. The Min., Met., & Mat. Soc.

Domenico, P.A., and Schwartz, F.W., 1998. *Physical and Chemical Hydrogeology*. New York, John Wiley & Sons, Inc., 506 pp.

Doyle, T.A., Davis, A., and Runnells, D.D., 1994. Predicting the environmental stability of treated copper smelter flue dust. *Appl. Geochem.*, v. 9, pp. 337–350.

Drever, J.I., 1988. *The Geochemistry of Natural Waters*, 2nd edn. Englewood Cliffs, New Jersey, Prentice-Hall, 437 pp.

Drez, P.E., 1988. Rock-water interactions between injected waste and host formation fluids and mineralogy during deep well injection. *EOS, Trans. Am. Geophys. Un.*, v. 69, no. 16, p. 350.

Driscoll, C.T., and Postek, K.M., 1996. The chemistry of aluminum in natural waters, Chapter 9, pp. 363–418, in G. Sposito, ed., *The Environmental Chemistry of Aluminum*, 2nd edn, CRC Press, Inc., Lewis Publishers, 464 pp.

Dulaney, A.R., 1989. The geochemistry of the N-aquifer system, Navajo and Hopi Indian Reservations, Northeastern Arizona. Unpublished Master thesis, Northern Arizona University, Flagstaff, Arizona, 209 pp.

Dzombak, D.A., and Hudson, R.J.M. 1995. The contributions of diffuse layer sorption and surface complexation, in C.P.H. Huang, C.R. O'Melia, and J.J. Morgan, eds., *Aquatic Chemistry. Interfacial and Interspecies Processes*, Advances in Chemistry Ser., vol. 244, Washington D.C., American Chemical Society, Chap. 3, pp. 59–93.

Dzombak, D.A., and Morel, F.M.M., 1990. *Surface Complexation Modeling: Hydrous Ferric Oxide*. New York, John Wiley & Sons, 393 pp.

Engesgaard, P., and Christensen, T. H., 1988. A review of chemical solute transport models, *Nordic Hydrol.*, v. 19, pp. 183–216.

Engi, M., 1992. Thermodynamic data for minerals: a critical assessment, Chapter 8, in G.D. Price and N.L. Ross, eds, *The Stability of Minerals*. London, Chapman and Hall, pp. 267–328.

Eugster, H. P., 1957. Heterogeneous reactions involving oxidation and reduction at high pressures and temperature. *J. Chem. Phys.* v. 26, pp. 1760–1775.

Eychaner, J. H., 1983. Geohydrology and effects of water use in the Black Mesa area, Navajo and Hopi Indian Reservations, Arizona. U.S. Geol. Survey Water Supply Paper, 2201, 26 pp.

Farley, K.J., Dzombak, D.A., and Morel, F.M.M., 1985. A surface precipitation model for the sorption of cations on metal oxides. *J. Colloid & Interface Sci.*, v. 106, pp. 226–242.

Feasby D.G., Blacnchette, M., Tremblay, G., and Sirois, L.L., 1991. The mine environment neutral drainage program. Paper presented at the Second International Conference on the Abatement of Acidic Drainage, Canadian Institute of Mine and Metallurgy, Montreal, Quebec, September 16–18, 1991.

Fetter, C.W., 1999. *Contaminant Hydrogeology*, 2nd edn. New York, Prentice-Hall Inc., 500 pp.

Freeze, R.A., and Cherry, J.A., 1979. *Groundwater*. New York, Prentice-Hall, 604 pp.

Fritz, S.J., 1994. A survey of charge-balance errors on published analyses of portable ground and surface water. *Ground Water*, v. 32, no. 4, pp. 539–546.

Gardiner, M.A. and Myers, J., 1992. Geochemical modeling of the deep injection well disposal of acid wastes into a Permian aquifer/aquitard system in Texas, USA, in Yousif K. Kharaka and Ann S. Maest, eds, *Proceedings – International Symposium on Water-Rock Interaction*, vol. 7, pp. 385–388.

Garrels, R. M., and Mackenzie, F. T., 1967. Origin of the chemical compositions of some springs and lakes, in R. F. Gould, ed., *Equilibrium Concepts in Natural Waters*, Advances in Chemistry Ser. vol. 67, Washington DC, American Chemical Society, pp. 222–242.

Garrels, R.M., and Thompson, M.E., 1962. A chemical model for sea water at 25 °C and one atmospheric pressure. *Am. J. Sci.* v. 260, pp. 57–66.

GeoTrans, Inc., 1987a. Tailings seepage control analysis for Bear Creek Uranium Company, Converse Co., Wyoming. Project No. 126.1.

GeoTrans, Inc., 1987b. A two-dimensional, finite-difference flow model simulating the effects of withdrawals to the N aquifer, Black Mesa area, Arizona. Submitted to the Office of Surface Mining on behalf of Peabody Western Coal Company.

Glynn, P., and Brown, J., 1996. Reactive transport modeling of acid metal-contaminated groundwater at a site with sparse spatial information, in P.C. Lichtner, C.I. Steefel, and E.H. Oelkers, eds, *Reactive Transport in Porous Media*, Review in Mineralogy. Washington D.C., Mineralogical Society of America, pp. 377–438.

Goldstein, J.I. *et al.*, 1992. *Scanning Electron Microscopy and X-ray Microanalysis*. New York, Plenum Press, 2nd edn, 820 pp.

Goodwin, H., 1962. Half-life of radiocarbon, *Nature*, v. 195, pp. 984–985.

Goolsby, D. A., 1971. Hydrogeochemical effects of injecting wastes into a limestone aquifer near Pensacola, Florida. *Ground Water*, v. 9, no. 1, pp. 13–17.

Grove D.B., and Stollenwerk, K. G., 1987. Chemical reactions simulated by groundwater quality models. *Water Resources Bull.*, v.23, no. 4, pp. 601–615.

Gunter, W.D., Perkins, E.H., and Hutcheon, I., 2000. Aquifer disposal of acid gases: modeling of water-rock reactions for trapping of acid wastes. *Appl. Geochem.*, v. 15, pp. 1085–1095.

Harshbarger, J.W., Repenning, C.A. and Irwin, J.H., 1957. Stratigraphy of the Uppermost Triassic and the Jurassic Rocks of the Navajo Country. U.S. Geol. Survey Prof. Paper 291, 71 pp.

Hedin, R. S., Narin, R. W., and Kleinmann, R. L., 1993. Passive treatment of coal mine drainage. Department of Interior, Bureau of Mines, Circular 9389, 35 pp.

Helfferich, F. G., 1989. The theory of precipitation/dissolution waves. *AIChE J.*, v. 35, no. 1, pp. 75–87.

Helgeson, H.C., 1968. Evaluation of irreversible reactions in geochemical processes involving minerals and aqueous solutions - I. Thermodynamic relations. *Geochim. Cosmochim. Acta*, v. 32, pp. 853–877.

Helgeson, H.C., 1969. Thermodynamics of hydrothermal systems at elevated temperatures and pressures. *Am. J. Sci.*, v. 267, pp. 729–804.

Helgeson, H.C., 1979. Mass transfer among minerals and hydrothermal solutions. in H. L. Barnes, ed.,*Geochemistry of Hydrothermal Ore Deposits*, 2nd edn, New York, John Wiley & Sons, pp. 568–610.

Helgeson, H.C., Brown, T.H., Nigrini, A., and Jones, T.A., 1970. Calculation of mass transfer in geochemical processes involving aqueous solutions. *Geochim. Cosmochim. Acta*, v. 34, pp. 369–592.

Hem, J.D., 1985. Study and interpretation of the chemical characteristics of natural water. U.S. Geol. Survey, Water-supply paper 2254, 263 pp.

Hemphill, C.P., Ruby, M.V., Beck, B.D., Schoof, R., Davis, A., and Bergstrom, P.D., 1991. The bioavailability of lead in mining wastes: Physical/chemical considerations. *Chem. Speciation & Bioavail.*, v. 3, pp. 135–148.

Hendershot, W.H., Courchesne, F., and Jeffries, D.S., 1996. Aluminum geochemistry at the catchment scale in watersheds influenced by acidic precipitation, in G. Sposito, ed., *The Environmental Chemistry of Aluminum*, 2nd edn, Chapter 10, pp. 419–449. CRC Press, Inc., Lewis Publishers, 464 pp.

Herbert, R.B. Jr., 1996. Metal retention by iron oxide precipitation from acidic groundwater water in Dalarna, Sweden. *Appl. Geochem.*, v. 11, pp. 229–235.

Hunter, K. S., Wang, Y., and van Cappellen, P., 1998. Kinetic modeling of microbially-driven redox chemistry of subsurface environment: coupling transport, microbial metabolism and geochemistry, *J. Hydrology*, v. 209, pp. 53–80.

Ineson, P.R., 1989. *Introduction to Practical Ore Microscopy*. New York, Longman/Wiley, 181 pp.

Jackson, M.L., 1985. *Soil Chemical Analyses – Advanced Course*, 2nd edn, 11th printing. Madison, Wisconsin, 53705; published by the author.

Jambor, J.L., and Blowes, D.W., 1994. *Short Course Handbook on Environmental Geochemistry of Sulfide Mine Wastes*. Waterloo, Ontario, Mineralogical Association of Canada, vol. 22.

Johnson, J.W., Knauss, K.G., Glassley, W.E., DeLong, L.D., and Thompson, A.F.B., 1998. Reactive transport modeling of plug-flow reactor experiments: quartz and tuff dissolution at 240°C *J. Hydrology*, v. 209, no. 1–4, August, pp. 81–111.

Johnson, J.W., Oelkers, E.H., and Helgeson, H.C., 1992. SUPCRT: A software package for calculating the standard molal thermodynamic properties of minerals, gases, aqueous species, and reactions as functions of temperature and pressure. *Computer & Geosci.*, v. 18, pp. 899–947.

Johnson, T.M. and DePaolo, D.J., 1996. Reaction-transport models for radiocarbon in groundwater: The effects of longitudinal dispersion and the use of Sr isotope ratios to correct for water-rock interactions. *Water Resources Res.*, v. 32, no. 7, pp. 2203–2212.

Keating, E.H., and Bahr, J.M., 1998. Using reactive solutes to constrain groundwater flow models at a site in northern Wisconsin. *Water Resources Res.* v. 34, no. 12, pp. 3561–3571.

Kerr, P.F., 1977. *Optical Mineralogy*. New York, McGraw-Hill, 492 pp.

Kharaka, Y.K., Ambats, G., Thordsen, J.J., and Davis, R.A., 1997. Deep well injection of brine from Paradox Valley, Colorado; potential major precipitation problems remediated by nanofiltration. *Water Resources Res.*, v. 33, no. 5, pp. 1013–1020.

King, T.V.V., (ed.), 1995. Environmental considerations of active and abandoned mine lands. *U.S. Geol. Survey Bull.*, v. 2220, 38 pp.

Knapp, R.A., 1989. Spatial and temporal scales of local equilibrium in dynamic fluid-rock systems. *Geochim. Cosmochim. Acta*, v. 53, pp. 1955–1964.

Knauss, K. G., and Wolery, T. J., 1986. Dependence of albite dissolution kinetics on *p*H and time at 25 °C and 70 °C. *Geochim. Cosmochim. Acta*, v. 52, pp. 43–53.

Kohler, M., Curtis, G.P., Kent, D.B., and Davis, J.A., 1996. Experimental investigation and modeling of uranium(VI) transport under variable chemical conditions. *Water Resources Res.*, v. 32, pp. 3539–3551.

Konikow, L.F., and Bredehoeft, J.D., 1992. Groundwater models cannot be validated. *Adv. Water Res.*, v. 15, pp. 75–83.

Kraynov, S.P., 1997. Thermodynamic models for groundwater chemical evolution versus real geochemical properties of groundwater: a review of capabilities, errors, and problems. *Geochem. Int.*, v. 35, pp. 639–655.

Lafon, G.M., Otten, G.A., and Bishop, A.M., 1992. Experimental determination of the calcite–dolomite equilibrium below 200°C; revised stabilities for dolomite and magnesite support near equilibrium dolomitization models. *Geol. Soc. Am. Abst. with Programs*, v. 24, pp. A210–A211.

Langmuir, D., 1997, *Aqueous Environmental Geochemistry*. New Jersey, Prentice-Hall, 600 pp.

Lasaga, A. C., 1998. *Kinetic Theory in the Earth Sciences*. Princeton University Press, 811 pp.

Li M. G., Jacob, C., and Comeau, G., 1996. Decommissioning of sulphuric acid-leached by rinsing. in *Proceedings of Tailings and Mine Waste '96*, Rotterdam, Balkema, pp. 295–304.

Lichtner, P.C., 1991. The quasi-stationary state approximation to coupled mass transport and fluid-rock reaction: local equilibrium revisited, in J. Ganguly, ed. *Diffusion, Atomic Ordering and Mass Transport*, Advances in Physical Geochemistry vol. 8, pp. 454–562.

Lichtner, P.C., 1993. Scaling properties of time-space kinetic mass transport equations and the local equilibrium limit. *Am. J. Sci.*, v. 293, pp. 257–296.

Lichtner, P.C., 1996. Continuum formulation of multicomponent-multiphase reactive transport, in P.C. Lichtner, C.I. Steefel, and E.H. Oelkers, eds, *Reactive Transport in Porous Media*, Washington, D.C., Review in Mineralogy, Mineralogical Society of America, pp. 1–82.

Littin, G.R., 1992. Results of ground-water, surface-water, and water-quality monitoring, Black Mesa area, northeastern Arizona-1990–91. U.S. Geol. Survey Water-Resources Invest., 92-4045, 32 pp.

Loeppert, R.H., and Inskeep, W.P., 1996. Methods of Soil Analysis. Soil Science Society of America, Madison, Wisconsin, pp. 639–664.

Maloszewski, P., and Zuber, A., 1991. Influence of matrix diffusion and exchange reactions on radiocarbon ages in fissured carbonate aquifers. *Water Resources Res.*, v. 27, pp. 1937–1945.

Mangold, D.C., and Tsang, C.F., 1991. A summary of subsurface hydrological and hydrochemical models. *Rev. Geophys.*, v. 29, pp. 51–79.

Mercer, J.W., and Faust, C. R., 1981. *Ground-water Modeling*. National Water Well Association, 60 pp.

Merino, E., 1975. Diagenesis in Tertiary sandstones from Kettleman North Dome, California. II. Interstitial solutions: distribution of aqueous species at 100°C and chemical relation to the diagenetic mineralogy. *Geochim. Cosmochim. Acta*, v. 39, pp. 1629–1645.

Merlivat, L., and Jouzel, J., 1979. Global climatic interpretation of the deuterium-oxygen-18 relationship for precipitation. *J. Geophys. Res.*, v. 84, pp. 5029–5033.

Miller, C. L., Landa, E. R., and Updegraff, D. M., 1988. Ecology aspects of microorganism inhabiting uranium mill tailings. *Microbiol. & Ecol.*. v. 14, pp. 141–155.

Morel, F.M.M., and Hering, J.G., 1993. *Principles and Applications of Aquatic Chemistry*. New York, John Wiley & Sons, 588 pp.

Morin, K. A., Cheery, J. A., Dave, N. K., Lim, T. P., and Vivyurka, A. J., 1988. Migration of acidic ground water seepage from uranium-tailings impoundments. 1. Field study and conceptual hydrogeochemical model. *J. Contaminant Hydrology*, v. 2, pp. 271–303.

Morrison, G.M.P., Batley, G.E., and Florence, T.M., 1989. Metal speciation and toxicity. *Chem. in Britain*, August, pp. 791–796.

Morrison, S.J., Spangler, R.R., and Tipathi, V.S., 1995a. Adsorption of uranium (VI) on amorphous ferric oxyhydroxide at high concentrations of dissolved carbon (IV) and sulfur (VI). *J. Contaminant Hydrology*, v. 17, pp. 333–346.

Morrison, S.J., Tipathi, V.S., and Spangler, R.R., 1995b. Coupled reaction/transport modeling of a chemical barrier for controlling uranium (VI) contamination in groundwater. *J. Contaminant Hydrology*, v. 17, pp. 347–363.

Murphy, E. M., and Schramke, J. A., 1998. Estimate of microbial rates in groundwater by geochemical modeling constrained with stable isotopes. *Geochim. Cosmochim. Acta*, v. 62, pp. 3395–3406.

Murphy, W. M., and Helgeson, H.C., 1988. Thermodynamics and kinetic constraints on reaction rates among minerals and aqueous solutions IV. Retrieval of rate constants and activation parameters for the hydrolysis constants for pyroxene, wollastonite, olivine, andalusite, quartz, and nepheline, *Am. J. Sci.*, v. 289, pp. 17–101.

National Research Council, 1990. *Ground Water Models - Scientific and Regulatory Applications.* Washington D.C., National Academic Press, 303 pp.

National Research Council, 1994. *Alternatives for Ground Water Cleanup.* Washington D.C., National Academy Press, 314 pp.

National Research Council, 2000. *Natural Attenuation for Groundwater Remediation.* Washington D.C., National Academy Press, 274 pp.

Neretnieks, I., 1981. Age dating of groundwater in fissured rock: Influence of water volume in micropores. *Water Resources Res.,* v. 17, no. 2, pp. 421–422.

Nordstrom, D.K., 1992. On the evaluation and application of geochemical models. Appendix 2, in *Proceedings of the 5th CEC Natural Analogue Working Group and Alligator Rivers Analogue Project,* and International Workshop, Toledo, Spain, October 5–19. EUR 15176 EN, pp. 375–385.

Nordstrom, D.K., and Ball, J.W., 1984. Chemical models, computer programs and metal complexation in natural water, in C.J. M. Kramer and J. Duniker, eds, *International Symposium on Trace Metal Complexation in Natural Water.* Martinus Nijhoff/Dr J.W. Junk Publishing Co., pp.149–169.

Nordstrom, D.K., and Munoz, J. L., 1986. *Geochemical Thermodynamics.* Palo Alto, Blackwell Scientific Publications, 477 pp.

Nordstrom, D.K., Plummer, L.N., Wigley, T.M.L., Wolery, T.J., and Ball, J.W., 1979. A comparison of computerized chemical models for equilibrium calculations in aqueous systems, in E.A. Jenne, ed., *Chemical Modeling in Aqueous Systems* Symposium Series 93. American Chemical Society, pp. 857–892.

Nordstrom, D.K., Plummer, L.N., Langmuir, D., Busenberg, E., May, H.M., Jones, B.F., and Parkhurst, D.L., 1990. Revised chemical equilibrium data for major water–mineral reactions and their limitations, in R.L. Bassett and D. Melchior, eds, *Chemical Modeling in Aqueous Systems II,* Symposium Series 416. American Chemical Society, pp. 398–413.

Nordstrom, D.K., Alpers, C.N., and Wright, W.G., 1996. Geochemical methods for estimating pre-mining and background water-quality conditions in mineralized areas. *Geol. Soc. Am. Abstr. with Programs,* vol. 28, no. 7, pp. A465.

Ogard and Kerrisk, 1984. Groundwater chemistry along flow paths between a proposed repository site and the accessible environment. Los Alamos National Laboratory, LA-10188-MS, UC-70, November, 1984.

Opitz, B.E., Dodson, M.E., and Serne, R.J., 1983. Uranium mill tailings neutralization: Contaminant complexation and tailing leaching studies. Pacific Northwest Laboratory, NUREG/CR-3906.

Oreskes, N., Schrader-Frechette, K., and Belitz, K., 1994. Verification, validation, and confirmation of numerical models in the Earth Sciences. *Science,* v. 263, pp. 641–646.

Palmer, D.A., and Drummond, S.E., 1986. Thermal decarboxylation of acetate; Part 1, The kinetics and mechanism of reaction in aqueous solution. *Geochim. Cosmochim. Acta*, v. 50, pp. 813–823.

Parkhurst, D.L., 1995. User's guide to PHREEQC – A computer program for speciation, reaction-path, advective-transport, and inverse geochemical modeling. U.S. Geol. Survey, Water-Resource Invest., pp. 95–4227.

Parkhurst, D.L., 1997. Geochemical mole-balance modeling with uncertain data. *Water Resources Res.*, v. 33, no. 8, pp. 1957–1970.

Parkhurst, D.L. and Appelo, A.A.J., 1999. User's guide to PHREEQC (version 2) – A computer program for speciation, batch-reaction, one dimensional transport, and inverse geochemical modeling. U.S. Geol. Survey, Water-resource Invest., pp. 99–4259.

Paschke, S.S., and van der Heijde, P., 1996. Overview of chemical modeling in groundwater and listing of available geochemical models. International Groundwater Modeling Center, Report GWMI–96-01, Golden, Co., Colorado School of Mines.

Penn, R. L., Zhu, C., Xu, H., and Veblen, D. R., 2001. Iron oxide coatings on sand grains from the Atlantic coastal plain: HRTEM characterization. *Geology*, v. 29, no. 9, pp. 843–846.

Perkins, E.H., 1992. Integration of intensive variable diagrams and fluid phase equilibrium with SOLMINEQ.88 pc.shell, in Y.K. Kharaka and A.S. Maest, eds, *Water-Rock Interaction*, Rotterdam, Balkema, pp. 1079–1081.

Phillips, F.M., Tansey, M.K., Peeters, L.A., Cheng, S., and Long, A., 1989. An isotopic investigation of ground water in the Central San Juan Basin, New Mexico: Carbon 14 dating as a basis for numerical flow modeling. *Water Resources Res.*, v. 25, pp. 2259–2273.

Phillips, O. M., 1991. *Flow and Reactions in Permeable Rocks*, New York, Cambridge University Press, 285 pp.

Plummer, L.N., 1985. Geochemical modeling: A comparison of forward and inverse methods, in B. Hitchon and E.I. Wallick eds, *First Canadian/American Conference on Hydrogeology, Practical Applications of Ground Water Geochemistry*. Worthington, OH, National Water Well Association.

Plummer, L.N., and Busenberg, E., 1982. The solubilities of calcite, aragonite and vaterite in CO_2–H_2O solutions between 0 and 90 °C, and an evaluation of the aqueous model for the system $CaCO_3$–CO_2–H_2O. *Geochim. Cosmochim. Acta*, v. 46, pp. 1011–1040.

Plummer, L.N., Parkhurst, D.L., and Thorstenson, D.C., 1983. Development of reaction models for groundwater systems, *Geochim. Cosmochim. Acta*, v. 47, pp. 665–685.

Plummer, L.N., Busby, J.F., Lee, R.W., and Hanshaw, B.B., 1990. Geochemical modeling of the Madison aquifer in parts of Montana, Wyoming, and South Dakota. *Water Resources Res.*, v. 26, pp. 1981–2914.

Plummer, L.N., Prestemon, E.C., and Parkhurst, D.L., 1991. An Interactive Code (NETPATH) for Modeling NET Geochemical Reactions along a flow PATH, Version 2.0. U.S. Geol. Survey Water-Resources Invest., 94–4169.

Plummer, L.N., Prestemon, E.C., and Parkhurst, D.L., 1992. NETPATH: An Interactive Code for interpreting NET geochemical reactions from chemical and isotopic data along a flow PATH, in Y. K. Kharaka and M. Maest eds, *Water-Rock Interaction*. Balkema, Rotterdam, pp. 239–242.

Plummer, L.N., Prestemon, E.C., and Parkhurst, D.L., 1994. An Interactive Code (NETPATH) for Modeling NET Geochemical Reactions along a flow PATH. U.S. Geol. Survey Water-Resources Invest., 91–4078.

Prigogine, I., and Defay, R., 1965. *Chemical Thermodynamics*. London, Longmans Green, 543 pp.

Raffensperger, J. P., 1996. Numerical simulation of sedimentary basin-scale hydrochemical processes. in M. Y. Corapcioglu, ed. *Advances in Porous Media*, Vol 3. Amsterdam, Elsevier, pp. 185–305

Raffensperger, J.P., and Garven, G., 1995. The formation of unconforming-type uranium ore deposit: 2. Coupled hydrochemical modeling. *Am. J. Sci.*, v. 295, pp. 581–636.

Reardon, E.J., 1981. K_d's – Can they be used to describe reversible ion sorption reactions in contaminant migration? *Ground Water*, v. 19, no. 3, pp. 279–286.

Reed, M.H., 1982. Calculation of multicomponent chemical equilibria and reaction processes in systems involving minerals, gases and an aqueous phase. *Geochim. Cosmochim. Acta*, v. 46, pp. 513–528.

Reed, S.J.B., 1996. *Electron Microprobe Analysis and Scanning Electron Microscopy in Geology*. Cambridge University Press, 201 pp.

Reisch, M., and Bearden, D. M., 1997. *Superfund Fact Book* on http://www.cnie.org/nle/waste-1.html. Updated March 3, 1997.

Rice J.M., and Ferry, J.M., 1982. Buffering infiltration, and the control of intense variables during metamorphism, in J.M. Ferry, ed. *Characterization of Metamorphism Through Mineral Equilibria*, Reviews in Mineralogy, vol. 10, Washington D.C., Mineralogical Society of America, pp. 263–326.

Ring, G.T., Turner, A.K., Aikin, A.R., and Jordan A.E., 1986. Geochemical aspects of artificial recharge by injection into the bedrock aquifers of the Denver groundwater basin. Abstracts with Program – Association of Engineering Geologists 29th Annual Meeting, p. 61.

Rittmann, B.E., and J.V. Van Briesen, 1996. Microbiological processes in reactive modeling, in P.C. Lichtner, C.I. Steefel, and E.H. Oelkers, eds, *Reactive Transport in Porous Media*, Review in Mineralogy, vol. 34, Washington D.C., Mineralogical Society of America, pp. 311–334.

Robins, R.G., 1990. The stability and solubility of ferric arsenate: an update. Paper presented at the Metallurgical Society Annual Meeting, Feb. 18–22, Anaheim, Ca., The Metallurgical Society, Warrendale, PA.

Rose, S. and Elliott, E. C., 2000. The effects of pH regulation upon the release of sulfate from ferric precipitation formed in acid mine drainage. *Appl. Geochem.*, v. 15, no. 1, pp. 27–34.

Rose, S. and Ghazi, A. M., 1997. Release of sorbed sulfate from iron oxyhydroxides precipitated from acid mine drainage associated with coal mining. *Environ. Sci. & Tech.*, v. 31, no. 7, pp. 2136–2140.

Ruby, M.V., Davis, A., Kempton, J.H., Drexler, J.W., and Bergstrom, P.D., 1992. Lead bioavailability: Dissolution kinetics under simulated gastric conditions. *Environ. Sci. & Tech.*, v. 26, pp. 1242–1248.

Runnells, D.D., Shepherd, T.A., and Angino, E.E., 1992. Determining natural background concentrations in mineralized areas. *Environ. Sci. & Tech.*, v. 26, pp. 2316–2323.

Sanford, W. E., 1997. Correcting for diffusion in carbon-14 dating of ground water. *Ground Water*, v. 35, no. 2, pp. 357–361.

Sanford, W. E., and Konikow, L. F., 1989. Simulation of calcite dissolution and porosity changes in seawater mixing zones in coastal aquifers. *Water Resources Res.*, v. 25, pp. 665–667.

Saunders, J.A., and Toran, L.E., 1995. Modeling of radionuclide and heavy metal sorption around low- and high-pH waste disposal sites at Oak Ridge, Tennessee. *Appl. Geochem.*, v. 10, pp. 673–684.

Sharp, W.N., and Gibbons, A.B., 1964. Geology and uranium deposits of the southern part of the Powder River Basin, Wyoming. *U.S. Geol. Survey Bull.*, v. 1147-D.

Smith, B.F.L. and Mitchell, B.D., 1987. Characterization of poorly ordered minerals by selective chemical methods, in M.J. Wilson, ed., *A Handbook of Determinative Methods in Clay Mineralogy*. New York, Chapman and Hall, pp. 275–294.

Smith, R.E., 1996. EPA mission research in support of hazardous waste injection 1986-1994, in J.A. Apps and C.-F. Tsang, eds, *Deep Injection Disposal of Hazardous and Industrial Waste*. Academic Press, pp. 9–24.

Smith, R.M., and Martell, A.E., 1998. *NIST Critically Selected Stability Constants of Metal Complexes Database, Version 5.0.* Gaithersburg, MD, U.S. Department of Commerce, National Institute of Standards and Technology, Standard Reference Data Program.

Smith, W.R., and Missen, R.W., 1982. *Chemical Reaction Equilibrium Analysis: Theory and Algorithms.* New York, John Wiley & Sons, 364 pp.

Sobek, A.A., Schuller, W.A., Freeman, J.R., and Smith R.M., 1978. Field and laboratory methods applicable to overburdens and minesoils. EPA manual No. EPA-600/2-78-054, section 3.2.3, pp. 47–50.

Solotto, B.V., *et al.*, 1966. Procedure for determination of mine waste acidity. Proceedings for the 154th National Meeting of the American Chemical Society, Chicago, Illinois.

Spence, J.C.H., 1988. *Experimental High Resolution Electron Microscopy*, Oxford, Oxford University Press.

Sposito, 1984, *The Surface Chemistry of Soils*, Oxford, Oxford University Press, 234 pp.

Srinivasan, P., and Mercer, J. W., 1987. Bio1D – one-dimensional model for comparison of biodegradation and adsorption processes in contaminant transport. Sterling, VA, GeoTrans, Inc.

Steefel, C., and MacQuarrie, K.T.B., 1996. Approaches to modeling of reactive transport in porous media, in P.C. Lichtner, C.I. Steefel, and E.H. Oelkers, eds, *Reactive Transport in Porous Media.* Review in Mineralogy, Washington D.C., Mineralogical Society of America, pp. 83–130.

Stollenwerk, K.G., 1991. Simulation of reactions affecting transport of constituents in the acidic plume, Pinal Creek Basin, Arizona, in J.G. Brown and B. Favor, eds, *Hydrology and Geochemistry of Aquifer and Stream Contamination Related to Acidic water in Pinal Creek Basin Near Globe, Arizona.* U.S. Geol. Survey Water Supply paper 2466, pp. 21–49.

Stollenwerk, K.G., 1994. Geochemical interactions between constituents in acidic groundwater and alluvium in aquifer near Globe, Arizona. *Appl. Geochem.*, v. 9, pp. 353–369.

Stumm, W., 1992. *Chemistry of the Solid-Water Interface.* New York, John Wiley & Sons, 428 pp.

Stumm, W., and Morgan, J., 1981. *Aquatic Chemistry Chemical Equilibria and Rates in Natural Waters*, New York, John Wiley & Sons, 428 pp.

Stumm, W. and Morgan, J.J., 1996. *Aquatic Chemistry: An Introduction Emphasizing Chemical Equilibria in Natural Waters*, 3rd edn. New York, John Wiley & Sons, 1022 pp.

Sudicky, E. A., and Frind, E. O., 1981. Carbon 14 dating of groundwater in confined aquifers: Implications of aquitard diffusion. *Water Resources Res.*, v. 17, no. 4, pp. 1060–1064.

Sumner, M. E., and Miller, W. P., 1996. Cation exchange capacity and exchange coefficients, in D. L. Sparks, ed. *Methods of Soil Analysis: Chemical Methods. Part 3.* Madison, Wisconsin, American Society of Agronomy, Inc., pp. 1021–1030.

Sverdrup, H.U., 1990. *The Kinetics of Base Cation Release Due to Chemical Weathering.* Lund, Lund University Press, 246 pp.

Sverjensky, D.A., 1993. Physical surface-complexation models for sorption at the mineral-water interface. *Nature*, v. 364, pp. 776–780.

Sverjensky, D.A., 1994. Zero-point-of-charge prediction from crystal chemistry and solvation theory. *Geochim. Cosmochim. Acta*, v. 58, pp. 3123–3129.

Sverjensky, D.A., and Sahai, N., 1996. Theoretical prediction of single-site surface-protonation equilibrium constants for oxides and silicates in water. *Geochim. Cosmochim. Acta*, v. 60, pp. 3773–3797.

Thomas, K., 1987. Summary of sorption measurements performed with Yucca Mountain, Nevada tuff samples and water from well J-13. Los Alamos National Laboratory, LA-10960-MS.

TRW (Environmental Safety Systems Inc.), 1999. Total System Performance Assessment–Site Recommendation Methods and Assumptions. TDR-MGR-MD-000001 REV 00 ICN 01. Available on the web site www.ymp.gov as of September 12, 2001.

Turner, D. R., 1995. A uniform approach to surface complexation modeling of radionuclide sorption. Southwest Research Institute, CNWRA 95–001.

Turner, D. R., and Pabalan, R. T., 1999. Abstraction of mechanistic sorption model results for performance assessment calculations at Yucca Mountain, Nevada. *Waste Management*, v. 19, no. 6, pp. 375–388.

U.S. EPA, 1985. Report to Congress on Injection of Hazardous Wastes. EPA 570/9-85-003.

U.S. EPA, 1990. Assessing the Geochemical Fate of Deep-well-injected Hazardous Wastes – Summaries of Recent Research. EPA/625/6-89/025b.

U.S. EPA, 1995. EPA's Composite Model for Leachate Migration with Transformation Products (EPACMTP) - Background Document for metals. EPA/SAB-EEC-95-010, v. 1.

U.S. EPA, 1997. Municipal Solid Waste: A Fact Book, version 4.0, EPA 530-C-97-001. US EPA, Washington, D.C., August 6, 1997.

U.S. EPA, 1999. The class V underground injection control study, in *Aquifer Recharge and Aquifer Storage and Recovery Wells* EPA/816-R-99-014u, vol. 21.

Valocchi, A. J., 1985. Validity of the local equilibrium assumption for the modeling sorbing solute transport through homogeneous soils. *Water Resources Res.*, v. 21, pp. 808–820.

VanBriesen J. M., and Rittmann, B. E., 1999. Modeling speciation effects on biodegradation in mixed metal/chelate systems. *Biodegradation* vol. 10, pp. 315–330.

van der Heijde, P.K.M., and Elnawawy, O.A., 1993. Compilation of groundwater models. U.S. Environmental Protection Agency, EPA/600/R-93/118. A report funded by EPA and prepared for the Holcomb Research Institute, Butler University, Indianapolis, Indiana 46208.

Van Zeggeren, F., and Storey, S.H., 1970. *The Computation of Chemical Equilibria.* London, Cambridge University Press, 176 pp.

Veblen, D.R., 1990. Transmission electron microscopy: Scattering processes, conventional microscopy, and high-resolution imaging, in I.D.R. Mackinnon and F.A. Mumpton, eds, *Electron and Optical Methods in Clay Science*, The Clay Minerals Society, CMS Workshop Lectures, 2, pp. 15–40.

Voss, C.I., and Wood, W.W., 1994. Synthesis of geochemical, isotopic and groundwater modeling analysis to explain regional flow in a coastal aquifer of southern Oahu, Hawaii, in *Proceedings of a Final Research Coordination Meeting on "Mathematical Models and Their Applications to Isotope Studies in Groundwater Hydrology"*, International Atomic Energy Agency in Vienna, Austria, June 1–4, 1993, pp. 147–178.

Wagman, D.D., Evans, W.H., Parker, V.B., Schumm, R.H., Halow, I., and Bailey, S.M., 1982. The NBS tables of chemical thermodynamic properties. *J. Phys. Chem. Ref. Data, II, Supplement no. 2.* Washington D.C., American Chemical Society.

Waite, T.D., Davis, J.A., Payne, T.E., Waychuna, G.A., and Xu, N., 1994. Uranium (VI) adsorption to ferrihydrite: Application of a surface complexation model. *Geochim. Cosmochim. Acta*, v. 58, pp. 5465–5478.

Walsh, M.P., 1983. Geochemical flow modeling. Unpublished Ph.D. dissertation, University of Texas, Austin, Texas.

Walsh, M.P., Bryant, S.L., Lake, L.W., and Schechter, R.S., 1984. Precipitation and dissolution of solids attending flow through porous media. *AIChE J.*, v. 30, no. 2, pp. 317–328.

Walter, A.L., Frind, E.O., Blowes, D.W., Ptacek, C.J., and Molson, J.W., 1994. Modeling of multicomponent reactive transport in groundwater. 1. Model development and evaluation. *Water Resources Res.*, v. 30, pp. 3137–3148.

Warner, D.L., and Lehr, J.H., 1977. An introduction of the technology of subsurface wastewater injection. EPA-600/2-77-240. Ada, Oklahoma, Robert S. Kerr Environmental Research laboratory.

Westall, J.C., Zachary, J.L., and Morel, F.M.M., 1976, MINEQL: A computer program for the calculation of chemical equilibrium composition of aqueous system. Technical Note 18, Department of Civil Engineering. Cambridge, MA, Massachusetts Institute of Technology, 91 pp.

White, A.F., and Chuma, N.J., 1987. Carbon and isotopic mass balance models of Oasis Valley-Fortymile Canyon groundwater basin, southern Nevada. *Water Resources Res.*, v. 23, no. 4, pp. 571–582.

White, A. F., Delany, J. M., and Smith, A., 1984. Groundwater contamination from an inactive uranium mill tailings pile 1. Application of a chemical mixing model. *Water Resources Res.*, v. 20, no. 11, pp. 1743–1752.

Whitworth, T.M., 1995. Hydrogeochemical computer modeling of proposed artificial recharge of the upper Santa Fe Group aquifer, Albuquerque, New Mexico. *New Mexico Geology*, v. 17, pp. 72–78.

Wickham, M., 1992. The geochemistry of surface water and groundwater interactions for selected Black Mesa drainage, Little Colorado river basin, Arizona. Unpublished Master thesis, University of Arizona, Tucson, Arizona, 249 pp.

Wigley, T.M., Plummer, L.N., and Pearson, F.J., 1978. Mass transfer and carbon isotope evolution in natural water systems. *Geochim. Cosmochim. Acta*, v. 42, pp. 1117–1139.

Willey, L.M., Kharaka, Y.K., Presser, T.S., Rapp, J.B., and Barnes, I., 1975, Short chain aliphatic acid anions in oil field waters and their contribution to the measured alkalinity. *Geochim. Cosmochim. Acta*, v. 39, pp. 1707–1711.

Williams, D.B. and Carter, C.B., 1996. *Transmission Electron Microscopy*. New York, Plenum Press, vol. I – IV.

Winterle J.R. and Murphy W.M., 1998. Time scales for dissolution of calcite fracture filling s and implications for saturated zone radionuclide transport at Yucca Mountain, Nevada, in *Proceedings of the Scientific Basis for Nuclear Waste Management XXII Symposium - Fall 1998 Meeting*. Warendale, PA, Material Research Society.

Wolery, T.J., 1992. EQ3/EQ6, a software package for geochemical modeling aqueous of systems, package overview and installation guide (version 7.0), Lawrence Livermore Laboratory, Report UCRL–MA–110662(1).

Yeh, G.T. and Tripathi, V.S., 1989. A critical evaluation of recent development of hydrogeochemical transport models of reactive multi-components. *Water Resources Res.*, v. 25, no. 1, pp. 93–108.

Yeh, G.T. and Tripathi, V.S., 1991. A model for simulating transport of reactive multispecies components: Model development and demonstration. *Water Resources Res.*, v. 27, no. 12, pp. 3075–3094.

Zheng, C., 1990. MT3D Reference Manual. Waterloo Hydrogeologic Inc.

Zhu, C., 2000. Estimate of recharge from radiocarbon dating of groundwater and numerical flow and transport modeling. *Water Resources Res.*, v. 36, no. 9, pp. 2607–2620.

Zhu, C., and Sverjensky, D.A., 1991. Partitioning of F-Cl-OH between minerals and hydrothermal fluids. *Geochim. Cosmochim. Acta*, v. 55, pp. 1837–1858.

Zhu, C., and Wille, J., 1997. Baseline of groundwater geochemistry at the Frenchman Flat, Nevada Test Site. Unpublished report to Department of Energy.

Zhu, C., and Murphy, W.M., 2000. On radiocarbon dating of ground water. *Ground Water*, v. 38, no. 6, pp. 802–804.

Zhu, C. and Burden, D. S., 2001. Mineralogical compositions of aquifer matrix as necessary initial conditions in reactive contaminant transport models. *J. Contaminant Hydrology* v. 51, no. 3–4, pp. 145–161.

Zhu, C., Sanders, S., and Rafal, M., 1993. Modeling coprecipitation reactions as the control of trace elements mobility in groundwater. *Geol. Soc. Am. Abstr. with Programs*, v. 25, no. 6, A-376.

Zhu, C., Xu, H., Ilton, E., Veblen, D., Henry, D., Tivey, M.K., and Thompson, G., 1994. TEM-AEM observations of high-Cl biotite and amphibole and possible petrological implications. *American Mineralogist*, v. 79, pp. 909–920.

Zhu, C., Waddell, R.K. Jr., Star, I., and Ostrander, M., 1998. Responses of groundwater in the Black Mesa, northeastern Arizona, to paleoclimate changes during late Pleistocene and Holocene. *Geology*, v. 26, pp. 27–130.

Zhu, C., Hu, F. Q., and Burden, D. S., 2001a, Multi-component reactive transport modeling of natural attenuation of an acid ground water plume at a uranium mill tailings site. *J. Contaminant Hydrology*, v. 52, pp. 85–108.

Zhu, C., Blum, A.E., Hedges, S., and White, C., 2001b. A tale of two rates: implications of the discrepancy between laboratory and field feldspar dissolution rates on geological carbon sequestration. *Geol. Soc. Am. Abstr. with Programs*, in press.

Zhu, C., Anderson, G. M., and Burden, D. S., 2002. Natural attenuation reactions at a uranium mill tailings site, western USA. *Ground Water*, v. 40, no. 1–2.

Index